CAD: Computational Concepts and Methods

Glen Mullineux

Kogan
Page

First published in 1986 by Kogan Page Ltd
120 Pentonville Road, London N1 9JN

British Library Cataloguing in Publication Data

Mullineux, G.
    CAD: computational concepts and methods.
    — (New technology modular series)
    1. Engineering design — Data processing
    I. Title   II. Series
    620'.00425'0285   TA174

    ISBN 1-85091-110-X

Printed and bound in Great Britain
by Biddles Ltd, Guildford

Contents

CHAPTER 1
Review of some basic ideas

Introduction

The purpose of this chapter is to review some of the standard notation and concepts that underlie the material to be presented later. These are the ideas of mathematics and of numerical mathematics. They are important from a computer-aided design (CAD) point of view because the only way in which we can persuade a computer system to deal with geometry for us is by reducing it to a set of numbers which can then be stored and manipulated.

Any manipulations that we carry out (such as finding the intersection between two pieces of geometry) are expressed as relations or equations between those numbers, and these equations then have to be solved. More often than not, no exact means are available for this solution and we need to turn regularly to approximate techniques. These often involve repeated corrections to an initial 'guess' solution; the improvements continuing until an acceptable level of accuracy is achieved. It is of course in this sort of area that the computer is powerful; it can perform these repetitious calculations very quickly.

In this chapter we look at simple coordinate geometry and at how vectors can be used in this area. The equations that arise when CAD geometry is manipulated are very often highly non-linear. Solution techniques however try to linearize them, so we remind the reader of some of the ideas of matrix manipulation. We then look at some iterative schemes for tackling non-linear problems.

Cartesian coordinates

In order to be able to handle geometric information within a computer system it is necessary to convert it to numerical data. The simplest way to carry out this representation is by use of Cartesian coordinates. Here points are described by giving their distances in relation to certain pre-selected lines called axes. If we are working in a single plane, two axes are used and these are normally chosen to be at right angles. For convenience the axes are usually labelled x and y and the point where they intersect is called the origin. Thus a typical point in a plane has two coordinates which are two numbers x and y being its distances from the origin in directions parallel to each axis; the point is at position (x, y). Figure 1.1 shows several points in a plane together with their appropriate

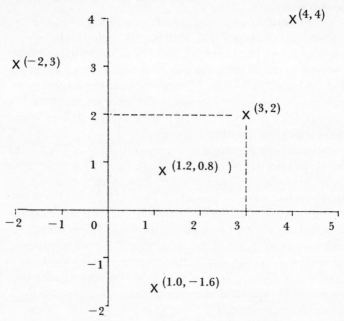

Figure 1.1 *Cartesian coordinate system*

coordinate positions. It will be noted that it is necessary to designate the parts of each axis on either side of the origin to be positive and negative and to call a coordinate value negative if the point lies towards the negative side of the appropriate axis. The position and orientation of the axes are essentially arbitrary. If we are attempting to represent a real object then it is often convenient to choose the origin at

some easily identified reference point and to have the axes
along lines of symmetry say.

If we move out of a plane and into general space we need
three axes and three coordinates to specify the position of
any point. That we need three and that just three will do
corresponds to the fact that we live in a three-dimensional
universe. The axes are again normally mutually perpendicular,
are labelled x, y and z, and intersect at a single point which is
the origin of the coordinate system. Figure 1.2 shows a point
in three-dimensional space and how its coordinates are

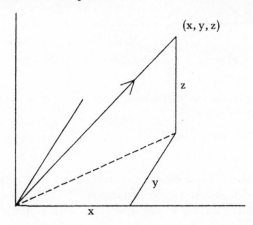

Figure 1.2 *Position and position vector of a point in space*

derived. The axes as drawn in the figure form a right handed
set; if we had reversed one of them (that is interchanged the
positive and negative parts of the axis) a left handed set
would have resulted. We use right handed sets of axes
throughout this book.

Various other means for assigning coordinates to positions
are possible. For example, when working with points in a
plane we could use polar coordinates. Again an origin is
selected and one coordinate of any point is its distance r
from the origin. An axis is also selected which is a line
starting from the origin (like the positive part of the x-axis).
The second coordinate for a point is the angle θ which the
line joining it to the origin makes with this axis. These
coordinates are shown in Figure 1.3. We can convert from

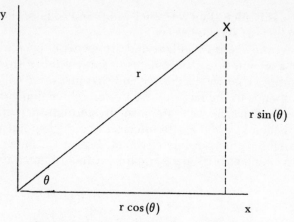

Figure 1.3 *Polar and Cartesian coordinates*

polar to Cartesian coordinates should we wish to do so using the following formulae:

$$x = r \cos(\theta) \qquad\qquad y = r \sin(\theta)$$
$$r = [x^2 + y^2]^{1/2} \qquad \theta = \arctan(y/x)$$

(Care should be taken when evaluating the arctan expression here to ensure that the correct angle is selected; for example if both x and y are negative the angle should be between 180 and 270°.)

Use of vectors

A vector is traditionally defined as a quantity which has both magnitude and direction. This is not altogether a very satisfactory definition. However we can see something of what it means by considering the position vector of a point in space. If we join the origin to the point then this line has magnitude (its length) and direction and sense (it goes from the origin to the point). We can mark this vector by an arrow as shown in Figure 1.2.

Three coordinates can be used to reference the point and these same three values can be used to denote the position vector of the point. In this case we usually write them as follows:

$$\begin{bmatrix} x \\ y \\ z \end{bmatrix}$$

and refer to this as a column vector whose components are x, y and z. Thus a vector is a group of components each of which is an ordinary number. When talking of vectors we usually refer to the ordinary numbers as scalars.

We can add and subtract vectors by adding and subtracting each of the individual components and we can multiply a vector by a scalar by multiplying each of its components by the scalar. Thus we can carry out the following sort of combination:

$$\begin{bmatrix} 1.5 \\ 2.0 \\ 0.5 \end{bmatrix} + 2 \begin{bmatrix} 1.2 \\ 0.0 \\ 0.5 \end{bmatrix} - \begin{bmatrix} 0.8 \\ 3.5 \\ 0.5 \end{bmatrix} = \begin{bmatrix} 3.1 \\ -1.5 \\ 1.0 \end{bmatrix} \qquad (1.1)$$

In order to avoid having to write down all the components each time we want to refer to a vector we often use a single letter and identify it as a vector by using bold print in this book. Usually **r** is used for the position vector of a point.

The length or magnitude of a vector can be obtained by adding the squares of all its components and taking the square root. The length is always non-negative. A vector of zero length is the zero vector **0**; its direction is unclear! A unit vector is one of length one. Unit vectors along the coordinates axes are defined by:

$$\mathbf{i} = \begin{bmatrix} 1 \\ 0 \\ 0 \end{bmatrix} \qquad \mathbf{j} = \begin{bmatrix} 0 \\ 1 \\ 0 \end{bmatrix} \qquad \mathbf{k} = \begin{bmatrix} 0 \\ 0 \\ 1 \end{bmatrix}$$

It is a simple matter to write any vector (with three components) in terms of these unit vectors. For example the result of Equation (1.1) can be expressed as:

$$3.1\mathbf{i} - 1.5\mathbf{j} + \mathbf{k}$$

Any vector in three-dimensional space can be expressed as this type of combination of these three unit vectors in a unique way. We say that **i**, **j**, and **k** form a basis.

If we move away from position vectors to free vectors we can ascribe geometric significance to the addition and subtraction of vectors. A free vector is one which does not necessarily emanate from the origin. Its components are the differences in x, y and z coordinates of the points at its two ends. The price we pay for this extension is that the free

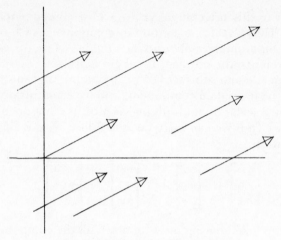

Figure 1.4 *Equivalent free vectors*

vector has no particular position in space. Figure 1.4 shows
in two dimensions many vectors all of which are equally well
$2i + j$.

Given two (free) vectors **a** and **b**, we can position them so
that they start from the same point. Then their sum is the
vector along the diagonal of the parallelogram (Figure 1.5).
Alternatively if the start of **b** is placed at the end of **a** then
their sum is the vector from the start of **a** to the end of **b**
(Figure 1.5 again). More generally the sum of a number of
vectors can be obtained geometrically by placing them
'nose-to-tail' and forming the vector from the start of the
first to the end of the last. The negative of a vector is formed

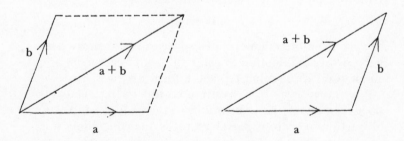

Figure 1.5 *Addition of vectors by the parallelogram and triangle rules*

by reversing its sense, so vectors can be subtracted in a similar graphical way.

There are two different products of two vectors that can be defined. Consider the vectors **a** and **b** as shown in Figure 1.6. The angle between them is θ and their lengths are

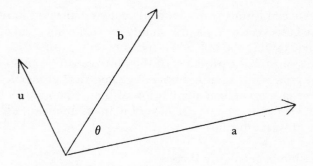

Figure 1.6 *Unit vector for the vector product*

a and b. The scalar product is denoted by **a.b** and is defined by:

$$\mathbf{a.b} = ab \cos(\theta)$$

The result is simply a scalar value. Thus for the unit vectors used previously we have:

$$\mathbf{i.i} = \mathbf{j.j} = \mathbf{k.k} = 1$$
$$\mathbf{i.j} = \mathbf{j.k} = \mathbf{k.i} = 0$$

Figure 1.6 shows **u** a unit vector perpendicular to both **a** and **b**. (It is defined in a right handed sense.) The vector product of the two vectors is defined by:

$$\mathbf{a} \wedge \mathbf{b} = ab \sin(\theta)\mathbf{u}$$

The result is a vector. For the unit vectors we have:

$$\mathbf{i} \wedge \mathbf{i} = \mathbf{j} \wedge \mathbf{j} = \mathbf{k} \wedge \mathbf{k} = 0$$
$$\mathbf{i} \wedge \mathbf{j} = \mathbf{k}$$
$$\mathbf{j} \wedge \mathbf{k} = \mathbf{i}$$
$$\mathbf{k} \wedge \mathbf{i} = \mathbf{j}$$

The surprising fact is that although we have defined these operations in terms of geometrical ideas, they obey most of the usual rules of multiplication. (The main exception is that

$b \wedge a$ is equal to $-a \wedge b$ and the minus sign is important.)
In particular we can remove brackets since for any vectors
a, **b** and **c**:

$$(a + b) . c = a . c + b . c$$

$$(a + b) \wedge c = a \wedge c + b \wedge c$$

Thus we can form the products of vectors expressed in terms
of the basis vectors **i**, **j** and **k** simply by removing brackets
and multiplying the individual terms together using the
expressions for the products of the unit vectors given above.
It is this easy interchange between geometrical properties of
vectors and algebraic manipulation of their components that
makes vectors such a powerful tool in many branches of
engineering mathematics.

Some simple geometric shapes

The use of coordinates allows us to describe points in three-
dimensional space as sets of numbers. When we describe more
complicated entities such as lines, curves and surfaces, we are
dealing with whole collections of points. If the entities are
reasonably simple there is usually some correspondingly
simple mathematical relationship between the coordinates of
all the points involved. We look briefly at a few examples.

In two dimensions, the simplest geometric entity which is
not a point is a straight line. Points which lie on a straight
line are often regarded as having coordinates which satisfy
the equation:

$$y = mx + c$$

where m is the gradient of the line. However for computational work, this formula has a drawback: it does not cope
well with the case when the line has infinite gradient.
A better form for the equation of a straight line is:

$$ax + by = c$$

In three dimensions the corresponding entity is the plane and
this has equation:

$$ax + by + cz = d$$

This form has the slight disadvantage that the parameters
a, b, c and d are not unique; we can multiply each of them by
any non-zero scalar value and the same plane is described.

We can take a step towards uniqueness by insisting that the sum of the squares of a, b and c be unity.

In order to describe a line in three-dimensional space we need two equations. In fact, more generally, one equation will normally define a surface in space; two equations will specify the curve where the two surfaces intersect (assuming they do); and three equations will normally yield the individual points where the three surfaces all meet. When we are dealing with planes the equations are not difficult to solve. For example, consider the pairs of equations:

$$x + 2y - z = 2$$
$$2x + z = 3$$

It is easily checked that for example the points $(1, 1, 1)$, $(3, -2, -3)$ and $(-1, 4, 5)$ all lie on both planes and hence on the line of intersection. In fact any point whose coordinates are:

$$x = 1 + 2t \qquad y = 1 - 3t \qquad z = 1 - 4t$$

for some value of t, lies on the line of intersection. Here we can regard t as a parameter. As t varies so the point described moves along the line.

An alternative form for expressing the equation of the above line is obtained by eliminating the parameter t and writing:

$$(x - 1)/2 = -(y - 1)/3 = -(z - 1)/4$$

which represents two equations connecting the three variables x, y and z. The reciprocals of the coefficients of these variables are the direction ratios of the line; in this case they are $2, -3$ and -4.

Lines and planes have the advantage that the equations that describe them are linear; they involve no products of coordinate values. Perhaps the most common non-linear geometric entity is the circle. In two dimensions, the equation of a circle of radius r and centre at the point (a, b) is:

$$(x - a)^2 + (y - b)^2 = r^2$$

This extends in the obvious way to the equation of a sphere in three-dimensional space.

Conic section curves are used frequently in certain design applications. These are curves in a plane and they comprise all those whose equation is of second degree. The typical equation for a conic section curve is:

$$ax^2 + 2bxy + cy^2 + dx + ey + f = 0$$

The value of $(b^2 - ac)$ determines the form of the conic. If it is negative, the curve is an ellipse; if zero, a parabola; and if positive, a hyperbola. Within these classifications there are also some degenerate cases. For example, if the coefficients are chosen correctly, a parabola may become a pair of straight lines.

Use of matrix notation

We introduced a vector as being a column of numbers. A matrix is simply a rectangular array of numbers. Thus:

$$A = \begin{bmatrix} 3.1 & 4.2 & 2.6 & -1.4 & 0.0 \\ -2.9 & 0.5 & 3.1 & 7.8 & 1.2 \\ 0.0 & 0.0 & 1.3 & 2.4 & 9.0 \\ 2.4 & -5.7 & 6.6 & 0.0 & 1.2 \end{bmatrix}$$

is a 4×5 matrix; it has four rows and five columns. We can regard matrices as a shorthand notation when we are dealing with a large number of simultaneous linear equations. For example, suppose we wish to find the single point of intersection of the following three planes:

$$\begin{aligned} x + y + z &= 2 \\ 2x + 3y + z &= 7 \\ 3x + y + 2z &= 3 \end{aligned} \qquad (1.2)$$

This can be written in terms of a 3×3 matrix A and column vectors **x** and **b** (which we can also regard as being 3×1 matrices) as follows:

$$Ax = b$$

where: $\qquad (1.3)$

$$A = \begin{bmatrix} 1 & 1 & 1 \\ 2 & 3 & 1 \\ 3 & 1 & 2 \end{bmatrix} \qquad x = \begin{bmatrix} x \\ y \\ z \end{bmatrix} \qquad b = \begin{bmatrix} 2 \\ 7 \\ 3 \end{bmatrix}$$

The expression Ax on the left of Equation (1.3) is regarded as a product of two matrices and we define the product in such a way that this matrix equation means precisely the same thing as Equation (1.2). Here is another advantage of matrix notation; we have written three equations as one and we find that properties of ordinary equations pass over to matrix equations and give insight into how they behave.

If A and B are matrices and B has the same number of rows as A has columns then we can form the product $C = AB$. If A is m x n and B is n x p then the product is m x p. The entry in row i, column j of the product is formed by multiplying together pairs of corresponding entries in row i of A and column j of B and adding the n resulting products. If a_{ik} denotes the entry in row i, column k of A and a similar notation is used for the other matrices then we have defined:

$$c_{ij} = a_{i1}b_{1j} + a_{i2}b_{2j} + \ldots + a_{in}b_{nj}$$

$$= \Sigma \, a_{ik}b_{kj}$$

The following is an example of a matrix product using numbers rather than symbols:

$$\begin{bmatrix} 1 & 2 & 1 & 3 \\ 2 & 0 & 0 & -1 \\ 1 & -1 & -2 & 2 \end{bmatrix} \begin{bmatrix} 1 & 0 \\ 1 & 2 \\ 0 & -1 \\ 2 & 1 \end{bmatrix} = \begin{bmatrix} 9 & 6 \\ 0 & -1 \\ 4 & 2 \end{bmatrix}$$

If we interchange the rows and columns of a matrix we obtain a new matrix called the transpose of the first. It is denoted by a T superscript; thus:

$$\begin{bmatrix} 9 & 6 \\ 0 & -1 \\ 4 & 2 \end{bmatrix}^T = \begin{bmatrix} 9 & 0 & 4 \\ 6 & -1 & 2 \end{bmatrix}$$

and if **a** is a column vector then its transpose $\mathbf{a}^T$ is a row vector. In particular if **b** is another column vector (of the same size), then the matrix product $\mathbf{a}^T\mathbf{b}$ is a 1 x 1 matrix. Its entry is easily seen to be the scalar product of the two vectors. For this reason it is common deliberately to confuse a 1 x 1 matrix with the number it contains and to write:

$$\mathbf{a}^T\mathbf{b} = \mathbf{a}.\mathbf{b}$$

The identity matrix takes the place of the number one in ordinary arithmetic. The identity is always a square matrix and there is one for any size. It has zeros as most of its entries except down its main diagonal where the entries are all unity. Thus the 4×4 identity looks like:

$$\begin{bmatrix} 1 & 0 & 0 & 0 \\ 0 & 1 & 0 & 0 \\ 0 & 0 & 1 & 0 \\ 0 & 0 & 0 & 1 \end{bmatrix}$$

We use I to denote an identity matrix (and assume its size is obvious from the context). It is straightforward to check that (provided the matrix sizes are compatible) for any matrix A we have:

$$AI = A \qquad \text{and} \qquad IA = A$$

If A is a square matrix and B is a second matrix of the same size such that AB and BA are both the identity matrix, then we say that A is invertible and that B is the inverse of A. The inverse is denoted by A^{-1}. For ordinary numbers the inverse of x is $1/x$ provided of course that x is non-zero. In the same way, not every square matrix has an inverse. A matrix is invertible if and only if its determinant is non-zero. (For more about determinants see for example Ayres, 1974.)

The standard linear equation

By this we mean the matrix-vector equation:

$$Ax = b$$

where the matrix A and the column vector **b** are both known and the column vector **x** is to be found. Equation (1.2), where the intersection of three planes is to be found, is an example of such a problem. In general there may not be any solution possible; the individual equations represented by the matrix formulation may be mutually incompatible. If A is an $m \times n$ matrix, then if m is larger than n there are more equations than unknown components of **x** and we would be lucky if this sort of incompatibility was not present. If m is smaller than n and a solution is possible, then we would expect several solutions, in fact an infinite number of solutions, to exist. This is like the case of finding the intersection of two planes. Here $m = 2$ and $n = 3$ and the infinite number of solutions represent all the points along the line of intersection.

Normally we have $m = n$ and the matrix A is square.

A unique solution is possible in this case if A is invertible. That solution is simply $A^{-1}b$. However it is usually a bad idea from a computational point of view to find the inverse of a matrix explicitly. It takes too much time. A better approach is to try to eliminate the components of x one at a time. This amounts to subtracting multiples of certain rows of A and of b away from other rows until A is made into an upper triangular matrix (that is it only has non-zeros on and above its main diagonal). Once this has been done the equation can be solved directly. This technique is Gauss elimination (see, for example, Burden *et al.*, 1978).

A variation on the Gauss method is LU decomposition. Here the matrix A is written as the product LU of a lower and an upper triangular matrix. This is not difficult to accomplish (see Burden *et al.*, 1978). In the case of the matrix arising from Equation (1.2) one possible decomposition is:

$$\begin{bmatrix} 1 & 1 & 1 \\ 2 & 3 & 1 \\ 3 & 1 & 2 \end{bmatrix} = \begin{bmatrix} 1 & 0 & 0 \\ 2 & 1 & 0 \\ 3 & -2 & 1 \end{bmatrix} \begin{bmatrix} 1 & 1 & 1 \\ 0 & 1 & -1 \\ 0 & 0 & -3 \end{bmatrix}$$

To complete the solution process, we solve the following two matrix equations:

$$Ly = b \quad \text{and} \quad Ux = y$$

As the matrices involved are triangular, these equations are straightforward to solve. For the former, the first row gives us the first component of y directly. This value put into the equation from the second row gives the second component; and so on. For the latter equation we need to work from the last row. This gives the last component of x; the next higher gives the preceding component; and so on again. With numbers in, these equations become:

$$\begin{bmatrix} 1 & 0 & 0 \\ 2 & 1 & 0 \\ 3 & -2 & 1 \end{bmatrix} \begin{bmatrix} 2 \\ 3 \\ 3 \end{bmatrix} = \begin{bmatrix} 2 \\ 7 \\ 3 \end{bmatrix}, \quad \begin{bmatrix} 1 & 1 & 1 \\ 0 & 1 & -1 \\ 0 & 0 & -3 \end{bmatrix} \begin{bmatrix} 1 \\ 2 \\ -1 \end{bmatrix} = \begin{bmatrix} 2 \\ 3 \\ 3 \end{bmatrix}$$

showing that the three planes defined by the parts of equation (1.2) intersect at the point $(1, 2, -1)$.

Sometimes our $m \times n$ matrix A is not square, we have m larger than n so that the problem is over-determined and an exact solution is impossible (or at least unlikely). In such cases we can ask for the best possible solution provided we can say what this means. An error vector **e** can be introduced as the difference $(Ax - b)$; the components of **e** would all be zero if the solution **x** were exact. The product $E = e^T e$ is the sum of the squares of the components of this vector and is used as a positive number measuring the size of the error. Again this number is zero if the solution is exact. The value E can be treated as a function of **x** and the point at which its value is minimal can be sought. Forming the partial derivatives with respect to the components of **x** and equating these all to zero shows that the minimum is achieved when:

$$A^T A x = A^T b$$

Effectively we have premultiplied both sides of the original linear equation by A^T and in so doing have made the matrix multiplier $A^T A$ of **x** on the left a square matrix. Thus we can begin to solve this new equation. If there is an exact solution to the original form, then the revised form certainly produces it. If not, then it generates a reasonable compromise solution. The way in which we set up this second form is the method of least squares, this name deriving from the fact that the sum of the squares of the components of the error vector is minimized. The solution obtained is the best possible one 'in the least squares sense'.

Non-linear equations

If we have one or more non-linear equations to solve, then the probability is that there is no (known) way to do so and obtain an exact solution. There are instances where this is possible. For example, if we want to find the intersection of a straight line and a circle, then it is possible to set this up as the solution of a quadratic equation and use the standard formula to write down the answer. However, this involves the use of a square root sign and from a computational point of view this cannot be evaluated immediately; some iterative scheme is necessary to carry out the work.

Iterative approaches are what we need to solve non-linear problems in general. They work by taking some initial guess to the required solution and then attempting to refine this until a solution of sufficient accuracy is obtained. While this approach works well in most cases, there are occasions when unexpected things happen and it is necessary for computer software adopting these techniques constantly to monitor how the sequences of approximations are behaving. In some cases bad behaviour can indicate that no solution is possible as would be the case for example if we were to attempt to find the intersection of two curves which did not in fact meet.

We begin by considering a single function of a single variable, $f(x)$. The problem is to find solutions of the equation $f(x) = 0$. Most simple non-linear problems can be put into this form. One approach is to try to rearrange this equation so that it becomes something like the following:

$$g(x) = x$$

For example if $f(x)$ were the polynomial:

$$x^3 - 4x + 1$$

(whose graph is shown in Figure 1.7), we might try the rearranged form given by:

$$(x^3 + 1)/4 = x$$

If our initial guess solution is x_0, then we substitute it into g and define $x_1 = g(x_0)$. Continuing in this way we obtain a sequence of values x_i where $x_{i+1} = g(x_i)$. If these values

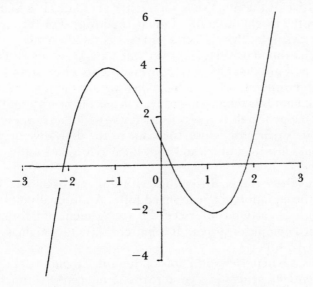

Figure 1.7 *A cubic polynomial function*

converge, then the limit reached is a solution to the equation. For the above rearrangement, using the two different starting points $x_0 = -2$ and $x_0 = 0$, the sequences obtained are as follows:

$$x_0 = -2.0000 \qquad x_0 = 0.0000$$
$$x_1 = -1.7500 \qquad x_1 = 0.2500$$
$$x_2 = -1.0898 \qquad x_2 = 0.2539$$
$$x_3 = -0.0736 \qquad x_3 = 0.2541$$
$$x_4 = 0.2499 \qquad x_4 = 0.2541$$
$$x_5 = 0.2539$$
$$x_6 = 0.2541$$
$$x_7 = 0.2541$$

It can be checked that 0.2541 is indeed a root of the original cubic polynomial. Note also that when we start at -2 the iterations take longer to converge as this start value is further away from the limit point. What is disappointing is that when we start at -2 we do not converge to the root which the graph of the function suggests is very close by. This is because the rearrangement we have chosen is not a good one for finding this particular root. Indeed there are rearrangements which do not permit convergence to any solution.

In fact a sufficient condition for convergence is that the derivative of the function g should have modulus less than one in the neighbourhood of the desired root. For the above example, $g'(x)$ is $3x^2/4$ which is certainly small enough around 0.2541. However when x is close to -2 the value is approximately 3, so convergence to a root in this neighbourhood is not possible. When we start at $x = -2$, the approximations tend to move away from the nearby root. They happen to drift towards the 0.2541 root and, once they are sufficiently close, the value of the derivative of g becomes less than unity and convergence to that root is assured.

Thus there can be difficulties with this approach once we have the equation in a rearranged form. Additionally, it is difficult to carry out the needed rearrangement of a formula within a computer system; it is better to use the original form of a function.

This is where Newton-Raphson iteration is important. It essentially generates a good rearrangement of the function for us automatically. If we are solving $f(x) = 0$, then we again define a sequence of values starting with some initial value x_0, and defining later ones by:

$$x_{i+1} = x_i - f(x_i)/f'(x_i)$$

where the prime denotes the derivative of the function. If we apply this to the cubic polynomial used previously then the following sequences are obtained:

x_0 =	-2.0000	x_0 =	0.0000
x_1 =	-2.1250	x_1 =	0.2500
x_2 =	-2.1150	x_2 =	0.2541
x_3 =	-2.1149	x_3 =	0.2541
x_4 =	-2.1149		

Beginning with the same starting values that we used before, we have now found two different roots of the cubic, namely -2.1149 and 0.2541. Note that in this case we have achieved convergence to a solution close (in some sense) to the initial guessed value. This is an advantage from a CAD point of view. For example, if we are interested in determining the intersection of two curves displayed on the graphics screen which in fact intersect at several points, it is possible to have the user indicate the approximate location of the desired intersection and then to use this as the start point for the iterations. We would not want our numerical method to start heading for some other intersection.

The Newton-Raphson method requires the evaluation of the derivative of the function. In some instances this is known exactly as an algebraic formula. If this is not the case, then the derivative value can be approximated using the following formula:

$$f'(x) = [f(x-h) + f(x+h)]/2h$$

where h is a suitably small value.

Both this formula and indeed the Newton-Raphson method itself can be obtained by consideration of the Taylor series for the function $f(x)$. If h is a small change in the value of x, then this series gives the value of $f(x+h)$ in terms of the values of f and its derivatives at x:

$$f(x + h) = f(x) + hf'(x) + h^2 f''(x)/2! + h^3 f'''(x)/3! + \dots$$

It is assumed here that all the derivates do indeed exist. If h is small and we ignore second and higher order terms in h, then we can deduce that:

$$f(x + h) = f(x) + hf'(x)$$

Thus, if x is an approximate solution to $f(x) = 0$ and $x + h$ is a better one, then we can take $f(x + h)$ to be zero, find that:

$$h = -f(x)/f'(x)$$

and deduce that the better solution is $[x - f(x)/f'(x)]$. This is precisely the Newton-Raphson scheme.

In practice, we often have to deal with several simultaneous non-linear equations in several unknowns. Suppose that there are m equations in n unknowns:

$$f_1(x_1, x_2, \ldots, x_n) = 0$$
$$f_2(x_1, x_2, \ldots, x_n) = 0$$
$$.$$
$$.$$
$$f_m(x_1, x_2, \ldots, x_n) = 0$$

We can write this more conveniently in terms of vector notation as follows:

$$f(x) = 0$$

Here **x** is the column vector comprising the n unknowns as its components and **f** and **0** are both column vectors of length m. We can again use the Taylor series idea. Here it gives to first order terms:

$$f(x + h) = f(x) + Jh$$

where J is the m x n Jacobian matrix whose entry in row i, column j is the partial derivative $\partial f_i / \partial x_j$. If $m = n$, then j is a square matrix. If we assume it is invertible, then we obtain the multi-dimensional form of Newton-Raphson iteration. This generates a sequence of vectors x_i where the first represents the initial guess and:

$$x_{i+1} = x_i - J^{-1} f(x_i)$$

As usual the explicit inversion of the matrix can be avoided by solving the matrix equation:

$$Jh = -f(x_i)$$

for **h** and then adding this on to x_i to obtain x_{i+1}.

This completes our initial look at numerical methods for

dealing with linear and non-linear problems. There has been no attempt to be exhaustive but only to present some of the key ideas. If one knows more about how a particular set of equations behaves or about precisely what information is available, it is possible to design methods which will work more efficiently than general purpose techniques. The interested reader will find many variations on the themes here presented in the literature (see, for example, Burden *et al.*, 1978; Scheid, 1968; Walsh, 1975).

CHAPTER 2

Simple datastructures and transformations

Introduction

It is possible to look at the representation of shapes within
a CAD system in two ways. First, we are interested in storing
the geometry and we look at this here in terms of certain of
the points or nodes that comprise the shape. Second, the
topology is important, that is how these nodes are joined
together to create the shapes we require.

Once we have defined some shapes it is necessary to be
able to manipulate them in order to refine the design being
produced. In this chapter we look at transformations of
shapes as translations, rotations and scalings. These are dealt
with in matrix terms. If individual components of a complex
assembly are created separately, it is possible to use these
types of transformation to put the assembly together. If the
assembly represents a mechanism, it is possible, by allowing
the transformations to vary, to simulate movement of the
mechanism and to check its operation.

Use of nodes

Basically a computer is a machine which can process data and
the data are essentially always numerical. Even if we think
we are processing text, as for example with a word processor,
the letters that are entered are converted into numbers by
some suitable code. Consequently, if we are going to manipulate graphical data with a computer we again need a means
for encoding it numerically. This means is provided by
Cartesian geometry. For geometry in a plane, each point
is defined by means of a pair of coordinate values with
reference to a predetermined set of axes (usually mutually
perpendicular). By storing and manipulating these two

numbers for every point we define, we can start to process graphical information.

In Cartesian geometry, lines and curves are considered as collections of points. The coordinates of each point are usually considered to be related by some equation. Now this means that each curve is regarded as being composed of an infinite number of points and clearly this approach is impractical for a computer with a finite memory size. However, most commonly used curves can be reconstructed from information about comparatively few points which lie on them. For example, a circle is uniquely defined by any three points which lie upon its circumference. In the same way a straight line is determined by any two points through which it passes.

We concentrate here upon shapes that are made up from straight line segments. To store and manipulate information about such segments we need only hold information about two points and it is most convenient to use the two end-points of the segment. We here refer to such points about which we hold information as nodes.

Node and edge lists

Thus we can consider simple two-dimensional shapes as being built up from nodes. We can store these nodes within the computer memory as a list of coordinates. We also need to record how they are interconnected. If two nodes are the end-points of a line segment we need to store this information. One approach to doing this is to create an edge list. This is essentially a list of pairs of nodes, the implication being that there is a line segment to be drawn between each pair. As an example, consider the rectangle of length 4 units and width 3 units shown in Figure 2.1. The coordinates of the corners are easily obtained and can be used to form the following list.

Node number	x coordinate	y coordinate
1	1.0	1.0
2	5.0	1.0
3	5.0	4.0
4	1.0	4.0

These nodes are joined in pairs to form the four edges of the rectangle, so we can produce the following edge list.

Edge number	First node	Second node
1	1	2
2	2	3
3	3	4
4	4	1

There is no particular significance about the order in which each pair of nodes is held. Ultimately, it may affect the way in which the shape is drawn out and it may be that some orderings are better than others from the point of view of searching through the edge list. Naturally we can add extra edges to the shape by adding to the edge list. For example, to add in a diagonal of the rectangle we could simply enter into the edge list a fifth entry indicating that nodes 1 and 3 are joined.

In this approach we have separated the geometry and the topology of the situation: the node list holds the geometric data and the edge list holds topological information, that is how the nodes are connected together. This means that if

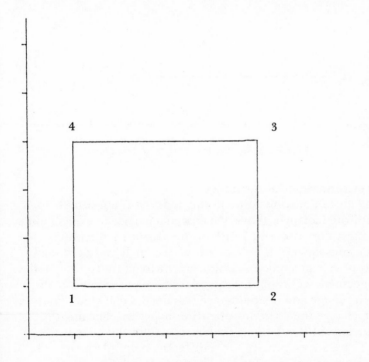

Figure 2.1 *A rectangular shape*

we move one of the nodes, then we produce a deformation of the original shape which preserves the interconnections. In the case of the rectangle above, if node number 3 is moved from position (5, 4) to position (7, 3) say, then the shape deforms to that of the quadrilateral shown in Figure 2.2.

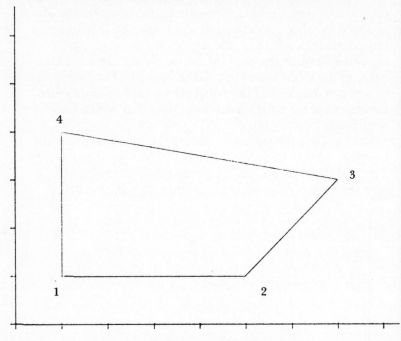

Figure 2.2 *Moving one corner of a rectangle*

Transformations of the nodes

The idea of moving nodes in the node list is important. With the rectangle above, we moved a particular node 2 units in the x-direction and 1 unit in the negative y-direction. As a consequence the shape of the rectangle was deformed. But if we had applied that movement to all four of the nodes concerned, we would have produced a repositioning of the entire shape and it would have remained a rectangle. Thus we can change the positions of entire shapes by changing the positions of the defining node held within the node list.

There are essentially three different types of changes of position or transformation. These are translation, rotation and scaling. Although we look first at the actions of these

30

upon points, they are transformations of shapes. The first
two are 'isometries', that is they preserve length of lines and
angles between lines. Scaling can be used to effect controlled
distortions of shapes.

A translation of a point in a plane is a movement of it by
a certain specified distance a in the x-direction and a
distance b in the y-direction. If we use the symbol := to
mean 'becomes equal to', then the effect of this translation
upon the point with coordinates (x, y) is as follows:

$$x := x + a \qquad (2.1)$$
$$y := y + b$$

At this stage we consider only rotations about the origin of
coordinates. Consider the situation in Figure 2.3. Here point

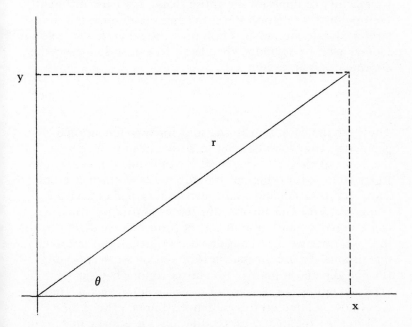

Figure 2.3 *Polar coordinates for a point*

(x, y) is at a distance r from the origin and the line joining it
to the origin makes an angle θ with the positive x-axis. Thus
the coordinates of the point can be expressed as:

$$x = r \cos(\theta)$$
$$y = r \sin(\theta)$$

31

If we use an anticlockwise angle ϕ of rotation about the origin, then the new coordinates of the point (x, y) are given by:

$$r \cos(\theta + \phi) \qquad \text{and} \qquad r \sin(\theta + \phi)$$

Hence, by expanding the sine and cosine of the sum of the two angles, we see that the change in the coordinates as a result of the rotation can be expressed as follows:

$$x := x \cos(\phi) - y \sin(\phi) \qquad (2.2)$$
$$y := y \sin(\phi) + x \cos(\phi)$$

By a scaling, we mean simply a multiplication of the x- and the y-coordinates of a point by suitable factors. If we use the same factor for both coordinates of all nodes in a shape, then we magnify or diminish the entire shape. If we use different factors, then we deform the shape and the result is that it is stretched more in one direction than the other. If the scale factors used are s_x and s_y then the effect of scaling upon the coordinates is given by:

$$x := s_x x \qquad (2.3)$$
$$y := s_y y$$

These are the basic transformations for two-dimensional geometry. They look to be quite different in form. From the point of view of seeing what is happening it is convenient to try to look for things in common between them. We can show that rotation and scaling can be regarded as having a common form. This involves the use of matrix notation. We can also write translation in matrix form, but we defer this to later on, once we have considered transformations in three dimensions. To use the matrix formulation we need to put the coordinates of points into matrix form. They can be arranged either as a row vector or as a column vector. Here we choose column vectors to denote points. (The choice does not matter greatly except that once it is made one needs to be consistent and it affects the order in which matrix multiplications are carried out.) Thus the point with coordinates (x, y) is written as the column vector:

$$\begin{bmatrix} x \\ y \end{bmatrix}$$

The action of rotation about the origin in a plane can then be

described as pre-multiplication of this column vector by a
2 × 2 matrix. Equation (2.2) can be written as:

$$\begin{bmatrix} x \\ y \end{bmatrix} := \begin{bmatrix} \cos(\phi) & -\sin(\phi) \\ \sin(\phi) & \cos(\phi) \end{bmatrix} \begin{bmatrix} x \\ y \end{bmatrix} \tag{2.4}$$

Scaling can also be written in this form and Equation (2.3)
becomes:

$$\begin{bmatrix} x \\ y \end{bmatrix} := \begin{bmatrix} s_x & 0 \\ 0 & s_y \end{bmatrix} \begin{bmatrix} x \\ y \end{bmatrix}$$

The advantage of trying to unify the description of transformations is that it can help to simplify the construction
of graphics programs. We can simply hold space for a
transformation matrix which can be filled in different ways
to describe various transformations. Then by pre-multiplying
it on to the appropriate coordinates in the node list the
transformation can be effected.

Matrix transformations can be combined if we so desire by
matrix multiplication. For example, the matrices to produce
a doubling of length in the x-direction and a rotation about
the origin of 60° are as follows:

$$\begin{bmatrix} 2.000 & 0.000 \\ 0.000 & 1.000 \end{bmatrix}, \begin{bmatrix} 0.500 & -0.866 \\ 0.866 & 0.500 \end{bmatrix}$$

The effect of the scaling followed by the rotation upon the
point (5, 1) is:

$$\begin{bmatrix} 0.500 & -0.866 \\ 0.866 & 0.500 \end{bmatrix} \begin{bmatrix} 2.000 & 0.000 \\ 0.000 & 1.000 \end{bmatrix} \begin{bmatrix} 5.0 \\ 1.0 \end{bmatrix} = \begin{bmatrix} 4.134 \\ 9.160 \end{bmatrix}$$

We could alternatively have first combined the two matrices
to obtain their product:

$$\begin{bmatrix} 0.500 & -0.866 \\ 0.866 & 0.500 \end{bmatrix} \begin{bmatrix} 2.000 & 0.000 \\ 0.000 & 1.000 \end{bmatrix} = \begin{bmatrix} 1.000 & -0.866 \\ 1.732 & 0.500 \end{bmatrix}$$

and then the new position of the point is given by:

$$\begin{bmatrix} 1.000 & -0.866 \\ 1.732 & 0.500 \end{bmatrix} \begin{bmatrix} 5.0 \\ 1.0 \end{bmatrix} = \begin{bmatrix} 4.134 \\ 9.160 \end{bmatrix}$$

which of course is the same result as before. The point

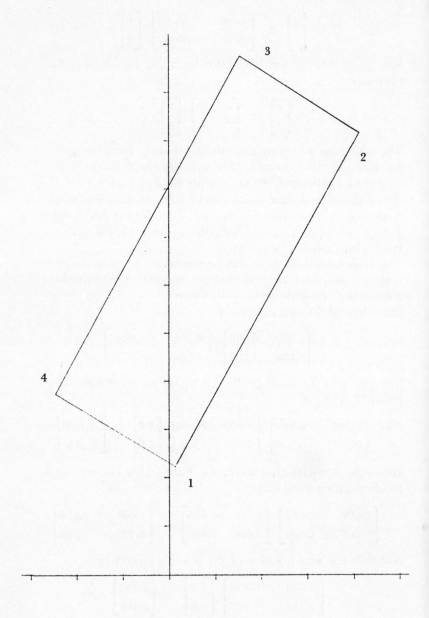

Figure 2.4 *Rotation of a rectangle*

transformed here is one of the nodes used to define a corner of the rectangle shown in Figure 2.1 and whose node list is given earlier. If the above scaling and rotation are applied to the four nodes in the list and the corresponding edge list used to produce the resulting shape, then the rectangle shown in Figure 2.4 is obtained.

Three-dimensional coordinates

So far we have only considered nodes in a plane. In the context of CAD this corresponds to those systems that are essentially only two-dimensional. They represent the 'electronic drawing board' in its closest sense. Using a two-dimensional system, a designer can reproduce all the usual constructional techniques that have developed with conventional drafting. However, the real power of CAD lies in the possibility of working in three dimensions and creating a single model of the component being designed. It is a model held within the computer memory. From this model all the usual views (such as plans and elevations) that are associated with a conventional design drawing can be created if they are required. As they are all derived from the single model there is greater confidence that they are mutually consistent.

Now although it offers far greater potential to the user, a three-dimensional CAD system does not differ greatly in the philosophy of its design from a two-dimensional one. One difference is that in the node lists we must store three coordinates to specify the position of each node; thus we are adding in a z-coordinate. The idea of the edge list remains the same; we still need only two nodes to define the ends of line segments. Naturally because there is more information inherent in the definition of a three-dimensional shape, the node and edge lists for shapes tend to be larger than in the two-dimensional case. Consider as an example the cuboidal shape shown in Figure 2.5. Its node and edge lists are as follows:

Node number	x-coordinate	y-coordinate	z-coordinate
1	1.0	1.0	0.0
2	5.0	1.0	0.0
3	5.0	4.0	0.0
4	1.0	4.0	0.0
5	1.0	1.0	3.0
6	5.0	1.0	3.0
7	5.0	4.0	3.0
8	1.0	4.0	3.0

Edge number	First node	Second node
1	1	2
2	2	3
3	3	4
4	4	1
5	5	6
6	6	7
7	7	8
8	8	5
9	1	5
10	2	6
11	3	7
12	4	8

The use of three coordinates presents us with a problem when we come to display the shape since this usually has to be done on a two-dimensional graphics device. With two dimensions, the data maps directly on to the device; with three we need to process it so that a usable image is produced. This requires view transformations to be introduced. We look at this aspect at a later stage. Our next step is to investigate

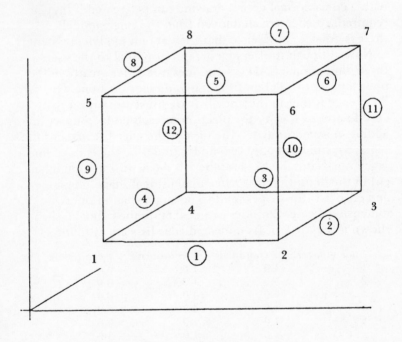

Figure 2.5 *Nodes and edges for a cuboid*

the possible transformations of three-dimensional shapes themselves.

Transformations in three dimensions

Again there are three basic transformations: translation, rotation and scaling. The first of these is achieved by adding (or subtracting) specified amounts from each of the x-, y- and z-coordinates of each node defining the shape. Rotation and scaling can be described by the use of matrices; now these are 3×3 matrices. To use them we need to write the coordinates of points as column vectors, in this case a vector with three components (a 3×1 matrix). Thus the point whose coordinates are (x, y, z) is represented by the column vector:

$$\begin{bmatrix} x \\ y \\ z \end{bmatrix}$$

Scaling is straightforward to render in matrix notation. If scale factors of s_x, s_y and s_z are introduced along the three coordinate axes, then the scaling transformation is equivalent to pre-multiplication by the diagonal matrix:

$$\begin{bmatrix} s_x & 0 & 0 \\ 0 & s_y & 0 \\ 0 & 0 & s_z \end{bmatrix}$$

Rotations in three dimensions are more complex. In two dimensions we need to specify a point about which the rotation is to take place; in three dimensions we need to define an axis of rotation. Some cases are easy. If the axis of rotation is the z-axis, then the transformation matrix can be obtained directly from the two-dimensional case in Equation (2.4); for a rotation through an angle γ the matrix is:

$$R_z = \begin{bmatrix} \cos(\gamma) & -\sin(\gamma) & 0 \\ \sin(\gamma) & \cos(\gamma) & 0 \\ 0 & 0 & 1 \end{bmatrix} \qquad (2.5)$$

A little thought shows that rotations through angles α and β about the x- and y-axes respectively can be described by the following matrices:

37

$$R_x = \begin{bmatrix} 1 & 0 & 0 \\ 0 & \cos(\alpha) & -\sin(\alpha) \\ 0 & \sin(\alpha) & \cos(\alpha) \end{bmatrix} \quad R_y = \begin{bmatrix} \cos(\beta) & 0 & -\sin(\beta) \\ 0 & 1 & 0 \\ \sin(\beta) & 0 & \cos(\beta) \end{bmatrix}$$

More generally, suppose that:

$$w = w_1 i + w_2 j + w_3 k$$

is a unit vector along the axis of a proposed rotation (so that w_1, w_2 and w_3 are the direction cosines of this axis). Then the transformation matrix R for a rotation of angle ϕ about this axis is given by:

$$\begin{bmatrix} c + (1-c)w_1{}^2 & -w_3 s + (1-c)w_1 w_2 & +w_2 s + (1-c)w_1 w_3 \\ +w_3 s + (1-c)w_1 w_2 & c + (1-c)w_2{}^2 & -w_1 s + (1-c)w_2 w_3 \\ -w_2 s + (1-c)w_1 w_3 & +w_1 s + (1-c)w_2 w_3 & c + (1-c)w_3 w_2 \end{bmatrix}$$

where $c = \cos(\phi)$ and $s = \sin(\phi)$.

We can obtain this matrix directly by looking at the geometry of the situation. However we can also obtain it in a different way. We first produce a transformation that takes the axis vector w onto the z-axis. Suppose that u and v are unit vectors which together with w form a right handed set of axes; in particular the vector product $u \wedge v$ of u and v is w. Then if P is the 3×3 matrix whose columns are formed by the components of u, v and w, then the transpose of P is a matrix transformation that takes w onto the z-axis. If we now rotate about the z-axis using R_z and then return the z-axis to the direction of w we have a compound transformation which fixes the direction w and produces a rotation about this axis. Thus the transformation matrix R given above is defined as:

$$R = P R_z P^T$$

We note in passing that if w is taken to be one of i, j or k, then the matrix R obtained simply reduces to R_x, R_y or R_z respectively.

Thus we have defined the transformations of scaling and rotation in terms of 3×3 matrices. This still leaves translation without a matrix formulation. It is possible to incorporate it, but it is necessary to introduce a change in approach to the

coordinates used to define points. This brings us to the idea
of homogeneous coordinates.

Homogeneous coordinates

Homogeneous coordinates are a means to allow us to unify
the description of the various transformations we wish to use.
They also allow a little greater flexibility in the way in which
points are described. We see this in more detail when we look
at curves. Homogeneous coordinates can also be used (from
a mathematical point of view) to deal in a convenient way
with 'points at infinity'.

So far we have used three numbers to describe positions in
space. We now introduce a fourth. Thus a point is represented
by four coordinates of the form (X, Y, Z, W). The last of
these, W, we refer to as the homogeneous coordinate or as
the 'weighting' of the point. In order to find which point is
being described, we need to form its Cartesian coordinates by
taking quotients. The point with the above coordinates
represents the point at Cartesian position (X/W, Y/W, Z/W).
In order to do this we need to assume that W is non-zero and
this we do throughout. In this way a single point in space can
be described by homogeneous coordinates in many ways. For
example, the following set of homogeneous coordinates all
refer to the point with Cartesian coordinates (1, 3, −2):

$$(1, 3, -2, 1) \qquad (2, 6, -4, 2) \qquad (-1.5, -4.5, 3.0, -1.5)$$

For a number of simple applications it is sufficient to take
the homogeneous coordinate to be unity. For this reason, it
is often not necessary explicitly to store the value and so save
memory space; it is reintroduced only when transformations
are to be performed. However, we will continue to use it
here so as to avoid any confusion.

Homogeneous transformation matrices

Thus homogeneous coordinates involve the use of four
numbers. As a consequence, transformations referred to these
coordinates are described by 4 × 4 matrices. As before we
need to represent the coordinates of a point as a column
vector; for example:

$$\begin{bmatrix} X \\ Y \\ Z \\ W \end{bmatrix}$$

We have previously looked at the 3×3 matrices used to perform rotations and scaling. If R is a 3×3 rotation matrix, then the corresponding 4×4 matrix for use with homogeneous coordinates is obtained by adding an extra fourth row and column to yield the following (in partitioned matrix notation):

$$\left[\begin{array}{ccc|c} & & & 0 \\ & R & & 0 \\ & & & 0 \\ \hline 0 & 0 & 0 & 1 \end{array}\right]$$

Suppose that $\mathbf{r} = x\mathbf{i} + y\mathbf{j} + z\mathbf{k}$ is the position vector of a typical point in space, so that its homogeneous coordinate representation is by the column vector $[\mathbf{wr}, w]^T$ for any non-zero value w. It is straightforward to check that pre-multiplication by the above matrix leads to the product:

$$\begin{bmatrix} \mathbf{wRr} \\ w \end{bmatrix}$$

This represents the point with Cartesian coordinates given by the column vector $\mathbf{Rr}$ which is of course the same result as applying R to the ordinary coordinates of the original point.

Scaling can be described by a 4×4 matrix in the same way as above. The 3×3 scaling matrix is used in place of R to produce in this case a diagonal matrix.

So far this is an extension of what we have done before. We now come to translations. Consider the 4×4 matrix:

$$\begin{bmatrix} 1 & 0 & 0 & a \\ 0 & 1 & 0 & b \\ 0 & 0 & 1 & c \\ 0 & 0 & 0 & 1 \end{bmatrix}$$

The result of pre-multiplying this onto the column vector $[wx, wy, wz, w]^T$ which represents the point (x, y, z) in space is to yield the column vector:

$$\begin{bmatrix} wx + wa \\ wy + wb \\ wz + wc \\ w \end{bmatrix}$$

This represents the point $(x + a, y + b, z + c)$, and thus the effect of the matrix is to produce a translation of the point. Thus we can now describe all transformations, including translations, by matrix means.

Note that in all three cases just described, the effect of the pre-multiplication of the various column vectors of positions is to leave the fourth coordinate unchanged. In the same way as remarked above about not storing the last coordinate, we can also hold the transformations as 3×4 rectangular matrices by ignoring the last row; this has the same form in all cases.

Naturally we can combine the 4×4 matrices to obtain more sophisticated transformations. As an example we produce one to describe a rotation through a right angle about a line through the point $(1, 2, 0)$ parallel to the z-axis. As a start consider the following two matrices:

$$T = \begin{bmatrix} 1 & 0 & 0 & -1 \\ 0 & 1 & 0 & -2 \\ 0 & 0 & 1 & 0 \\ 0 & 0 & 0 & 1 \end{bmatrix} \qquad R = \begin{bmatrix} 0 & -1 & 0 & 0 \\ 1 & 0 & 0 & 0 \\ 0 & 0 & 1 & 0 \\ 0 & 0 & 0 & 1 \end{bmatrix}$$

These represent respectively a translation of the point $(1, 2, 0)$ to the origin and a rotation through a right angle about the z-axis itself. The inverse of the translation T is obtained in this case by changing the sign of the first three entries in the last column. Thus we obtain the following which represents a translation of the origin to $(1, 2, 0)$:

$$T^{-1} = \begin{bmatrix} 1 & 0 & 0 & 1 \\ 0 & 1 & 0 & 2 \\ 0 & 0 & 1 & 0 \\ 0 & 0 & 0 & 1 \end{bmatrix}$$

Now by taking the required axis onto the z-axis by using T, performing the rotation about the z-axis and then moving it back to a line through $(1, 2, 0)$ we achieve the desired rotation. It has matrix:

$$T^{-1}RT = \begin{bmatrix} 0 & -1 & 0 & 3 \\ 1 & 0 & 0 & 1 \\ 0 & 0 & 1 & 0 \\ 0 & 0 & 0 & 1 \end{bmatrix}$$

41

We check this by looking at its effect upon the point $(3, 2, 1)$. The required matrix product is:

$$\begin{bmatrix} 0 & -1 & 0 & 3 \\ 1 & 0 & 0 & 1 \\ 0 & 0 & 1 & 0 \\ 0 & 0 & 0 & 1 \end{bmatrix} \begin{bmatrix} 3 \\ 2 \\ 1 \\ 1 \end{bmatrix} = \begin{bmatrix} 1 \\ 4 \\ 1 \\ 1 \end{bmatrix}$$

Thus the z-coordinate remains unchanged as we would expect and the effects upon the x- and y-coordinates can also be verified (cf. Figure 2.6).

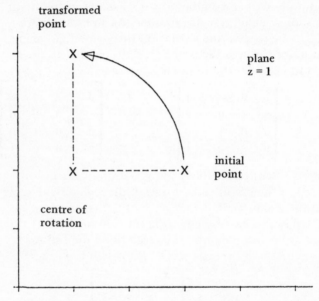

Figure 2.6 *Rotation about an axis parallel to the z-axis*

View transformations

We have been considering transformations of three-dimensional shapes. This is fine as far as transforming the data internal to a CAD system is concerned. Eventually however the information held needs to be presented to the user. The vast majority of graphic output devices are essentially two-dimensional displays. These are mainly graphics visual display terminals and plotters. (Some work on genuine three-dimensional displays has been carried out but the resulting devices are very expensive and two-dimensional

42

displays are going to be the predominant ones for a long time to come.) Consequently, we need to render the data about the nodes and shapes held into a form for two-dimensional output.

The simplest way to do this is merely to transform the data to ordinary three-dimensional Cartesian coordinates and throw away the third of these. This leaves two coordinates which can then be sent for plotting to the output device. Normally some form of final scaling will need to be performed in order to ensure that the output will fit well onto the device. This requires us to know the maximum and minimum x and y values that are to be plotted. Suppose these are x_{min}, y_{min}, x_{max} and y_{max}. We use the following to denote the maximum and minimum values to be plotted on the output device in device coordinates: Dx_{min}, Dy_{min}, Dx_{max} and Dy_{max}. These could be the extremes of the values that are allowable for plotting on the output device, or, by taking them inside this region, we can arrange to plot within a 'window' on the device. In this way perhaps several plots can be shown simultaneously on the same device. With this notation the largest possible scalings in the x- and y-directions are given by:

$$s_x = (Dx_{max} - Dx_{min})/(x_{max} - x_{min})$$
$$s_y = (Dy_{max} - Dy_{min})/(y_{max} - y_{min})$$

It may be desirable to take lower values than these on occasions. For example, to retain equal scaling in x and y the smaller of the two values should be taken. With the actual values to be used we define:

$$m_x = [(Dx_{max} - Dx_{min}) - s_x(x_{max} - x_{min})]/2$$
$$m_y = [(Dy_{max} - Dy_{min}) - s_y(y_{max} - y_{min})]/2$$

These represent margins at the sides of the window which will not be used for plotting. Then the required scaling of values x and y is as follows to obtain values x_{plot} and y_{plot} which are the ones to be plotted:

$$x_{plot} = Dx_{min} + m_x + s_x(x - x_{min})$$
$$y_{plot} = Dy_{min} + m_y + s_y(y - y_{min})$$

It may be that we wish to use a large scaling factor in order to magnify some part of a component held within the system. In such a case the plot would extend beyond the bounds of the plotting region unless some special precautions were taken. This is usually in the form of a 'clipping' algorithm which works to eliminate parts of the plot outside the window. Some output devices have software built into them that allows this to be done automatically. In other cases it is necessary to incorporate this software into the CAD system. The action of clipping is relatively simple and is illustrated in Figure 2.7. It is assumed that we are plotting straight line

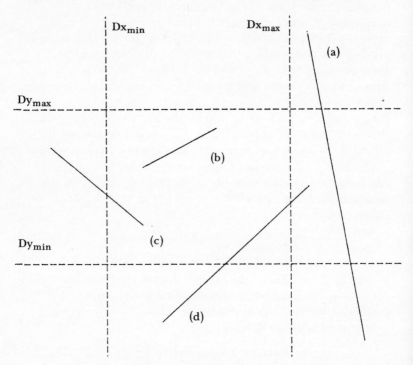

Figure 2.7 *Some cases in clipping*

segments and it is necessary to check each one against the bounds of the window used. This is done by looking at the end-points of the segment and seeing where they lie relative to the nine areas shown in the figure. The central area is the plotting window itself. The figure shows a number of segments and these would be dealt with as follows:

(a) both the end-points lie to one side (the right) of the window, so the segment is ignored entirely;

44

(b) both end-points lie within the window, so the whole segment is plotted;

(c) one end-point lies inside the window, the other outside, so the intersection of the segment with the window boundary must be found and only a portion of the segment plotted;

(d) both end-points lie outside the window but a part of the segment lies within it; in this case both intersections of the segment with the window boundary must be found and the appropriate portion plotted.

An algorithm to perform clipping is sketched in Table 2.1 (see also Foley and Van Dam, 1982; Newman and Sproull, 1979).

So far we have simply ignored the presence of the z-coordinate. It is possible to take account of this in two ways. Some output devices support the means of drawing lines in varying shades of the plotting colour. Thus we could use the third coordinate to establish which line segments (or parts of segments) are further away than others and draw these more faintly to give the impression of depth in the plot.

An alternative approach is to introduce perspective into the output plot. Here the z-coordinate is again used to indicate a distance away from the observer and is incorporated into a perspective transformation. There are several ways of defining such transformations. They can be rendered in matrix form. The simplest is perhaps the following:

$$\begin{bmatrix} 1 & 0 & 0 & 0 \\ 0 & 1 & 0 & 0 \\ 0 & 0 & 0 & 0 \\ 0 & 0 & -1/d & 1 \end{bmatrix}$$

Here the eye can be imagined at position (0, 0, d) along the z-axis and the view is projected onto the xy-plane. The point with homogeneous coordinates $[x, y, z, 1]^T$ is transformed to position $[xd/(d-z), yd/(d-z)]$. Note that if d is large (or infinite), then the matrix becomes diagonal and amounts to ignoring the z-coordinate of the point to be plotted. This type of transformation needs to be applied after the correct viewing direction has been set up as we next describe.

Finally, as an aid to viewing the entire shape of a component or part of it, the user of the CAD system can be given access to the basic transformations of translation, rotation and scaling and apply them to the complete shape. Translation here is equivalent to moving the plot across the

```
PROCEDURE Clip( VAR X1, Y1, X2, Y2 : REAL;
                   Xmin, Ymin, Xmax, Ymax : REAL;
               VAR Accept : BOOLEAN );

VAR    Out1, Out2 : ARRAY[1..4] OF BOOLEAN;
       Swap, Reject, Finish : BOOLEAN;
       I : INTEGER;

       PROCEDURE Do_Swap;
       VAR    Temp : REAL;
              Ltemp : BOOLEAN;
              I : INTEGER;
       BEGIN
           Temp:=X1;   X1:=X2;   X2:=Temp;
           Temp:=Y1;   Y1:=Y2;   Y2:=Temp;
           FOR I:=1 TO 4 DO
           BEGIN
               Ltemp:=Out1[I];  Out1[I]:=Out2[I];  Out2[I]:=Ltemp;
           END;
           Swap := NOT Swap;
       END;

BEGIN
    Accept:=FALSE;      Reject:=FALSE;
    Finish:=FALSE       Swap:=FALSE;

    REPEAT
    [Check against boundaries]
        Out1[1] := Y1>Ymax;    Out2[1] := Y2>Ymax;
        Out1[2] := Y1<Ymin;    Out2[2] := Y2<Ymax;
        Out1[3] := X1>Xmax;    Out2[3] := X2>Xmax;
        Out1[4] := X1<Xmin;    Out2[4] := X2<Xmin;

    FOR I := 1 TO 4 DO
        IF Out1[I] AND Out2[I] THEN  Reject:=TRUE;

    IF Reject
    THEN  Finish:=TRUE  [Segment is totally outside]
    ELSE  BEGIN
        Accept:=TRUE;
        FOR I := 1 TO 4 DO
            IF Out1[I] OR Out2[I] THEN  Accept:=FALSE;

        IF Accept
        THEN  BEGIN  [Segment now totally inside]
            Finish:=TRUE;
            IF Swap THEN  Do_Swap;
            END
        ELSE  BEGIN
            IF NOT(Out1[1] OR Out1[2] OR Out1[3] OR Out1[4])
            THEN Do_Swap;  [Ensure first point is outside]

            IF Out1[1]
            THEN  BEGIN [Clip to upper boundary]
                X1 := X1 + (X2-X1)*(Ymax-Y1)/(Y2-Y1);
                Y1 := Ymax;
                END
            ELSE
            IF Out1[2]
            THEN  BEGIN [Clip to lower boundary]
                X1 := X1 + (X2-X1)*(Ymin-Y1)/(Y2-Y1);
                Y1 := Ymin;
                END
            ELSE
            IF Out1[3]
            THEN  BEGIN [Clip to right boundary]
                Y1 := Y1 + (Y2-Y1)*(Xmax-X1)/(X2-X1);
                X1 := Xmax;
                END
            ELSE
            IF Out1[4]
            THEN  BEGIN [Clip to left boundary]
                Y1 := Y1 + (Y2-Y1)*(Xmin-X1)/(X2-X1);
                X1 := Xmin;
                END;
            END;
        END;
    UNTIL Finish;
END;
```

Table 2.1 *An algorithm to perform clipping*

display of the output device (panning); scaling can be used to 'zoom' into and away from the component (for example to enlarge a small part of it for closer examination); and rotation can be used to look from different view points. There are several ways of inputting the desired transformation to be applied. The simplest is to restrict the user to rotations about the current x-, y- and z-axes and so use the R_x, R_y and R_z matrices introduced earlier. This however can easily become confusing to the user, especially after several such rotations have been performed and it is no longer clear how the current axes relate to the original ones.

An alternative approach is to use the idea of a view vector. This is a direction in space along which the eye is imagined to look. It can be specified by the user as three values. These can be regarded as the coordinates of a position in space at which the eye is imagined to be looking straight at the origin. To effect the appropriate transformation, it is necessary to perform rotation to bring a unit vector **w** in this direction parallel to the z-axis. If d is defined to be the value:

$$d = (1 - w_3^2)^{1/2} = (w_1^2 + w_2^2)^{1/2}$$

and d is non-zero, then this rotation is performed by the following transformation matrix. (It is arranged that this causes the original z-axis to appear vertical on the resulting plot.)

$$\begin{bmatrix} -w_2/d & w_1/d & 0 & 0 \\ -w_1 w_3/d & -w_2 w_3/d & d & 0 \\ w_1 & w_2 & w_3 & 0 \\ 0 & 0 & 0 & 1 \end{bmatrix}$$

If d is zero, then w_3 is $+1$ or -1 and the following transformation matrix has been found to be suitable.

$$\begin{bmatrix} w_3 & 0 & 0 & 0 \\ 0 & 1 & 0 & 0 \\ 0 & 0 & w_3 & 0 \\ 0 & 0 & 0 & 1 \end{bmatrix}$$

Instead of specifying the viewing direction in terms of a vector, we can use the so-called observer angles. These are shown in Figure 2.8. The first angle θ represents a rotation about the current vertical axis. This is followed by a rotation

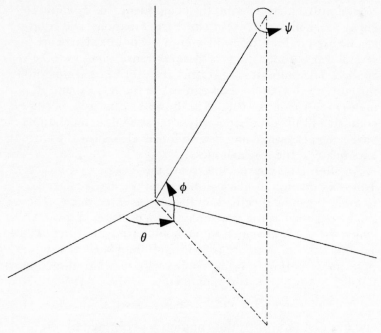

Figure 2.8 *Observer coordinate system*

through an angle ϕ about a line in the horizontal plane as
shown in the figure. These two angles are used to define the
position of the observer and his line of sight is from this
position to the origin. Thus in the previous notation we have
for the unit vector **w**:

$$w_1 = \cos\phi\cos\theta$$
$$w_2 = \cos\phi\sin\theta$$
$$w_3 = \sin\phi$$

and the transformation matrix is given by:

$$\begin{bmatrix} -\sin\theta & \cos\theta & 0 & 0 \\ -\sin\phi\cos\theta & -\sin\phi\sin\theta & \cos\phi & 0 \\ \cos\phi\cos\theta & \cos\phi\sin\theta & \sin\phi & 0 \\ 0 & 0 & 0 & 1 \end{bmatrix}$$

If, as suggested in Figure 2.8, we also allow a rotation
through an angle ψ about the viewing direction, then the

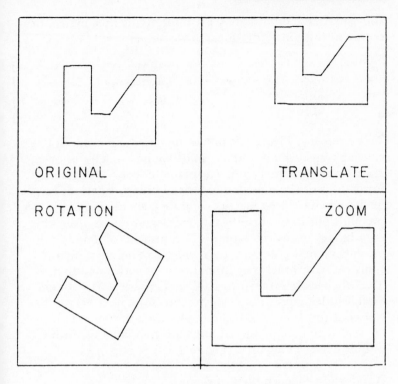

Figure 2.9 *Examples of simple transformations*

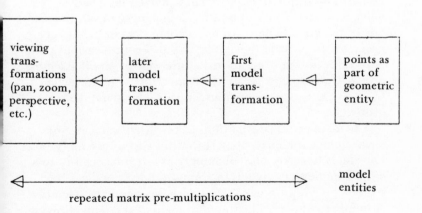

Figure 2.10 *Order of model and viewing transformations*

49

above matrix is modified by pre-multiplying by a matrix to represent this extra rotation and we obtain the following:

$$\begin{bmatrix} -\cos\psi\,\sin\theta + \sin\psi\,\sin\phi\,\cos\theta & \cos\psi\,\cos\theta + \sin\psi\,\sin\phi\,\sin\theta & -\sin\psi\,\cos\phi & 0 \\ -\sin\psi\,\sin\theta - \cos\psi\,\sin\phi\,\cos\theta & \sin\psi\,\cos\theta - \cos\psi\,\sin\phi\,\sin\theta & \cos\psi\,\cos\phi & 0 \\ \cos\phi\,\cos\theta & \cos\phi\,\sin\theta & \sin\phi & 0 \\ 0 & 0 & 0 & 1 \end{bmatrix}$$

As a summary, Figure 2.9 shows the types and order of viewing transformation that can be applied to a point or collection of points forming a complete shape. Both transformations of the models and viewing transformations can be described by 4×4 matrices and they are effected by pre-multiplication onto the column vector representing the coordinates of the relevant position (usually of a node); this produces a new column vector which can be subsequently transformed. Model transformations are performed first. Then the viewing transforms are applied; we start with one to establish the appropriate viewing direction, next perspective is used if this is required, and finally the values to be plotted are scaled and clipped to fit onto the graphics area of the output device being used (Figure 2.10).

Application to assemblies of shapes

Very often components that are designed using CAD are composed of a number of well-defined parts. These may be designed by different people. They may be designed to move with respect to each other as in a mechanism. Indeed one of the great benefits of using computers in the design process is to provide the ability to put together various parts 'electronically' in order to make sure that they will assemble properly and so correct some of the errors without having to go to the expense of manufacturing a large number of prototypes.

The ideas of transformations can be used to create the environment for a user to perform this sort of assembly task. The aim is to relate one subcomponent or subassembly to a more complex one by a transformation. Eventually a tree structure is built up which describes the complete assembly. A simple example is illustrated in Figure 2.11. This shows a schematic picture of a mechanical vice holding a rectangular block. The sliding part of this vice has been defined in terms of three sub-parts: the handle, the shank and the moving jaw.

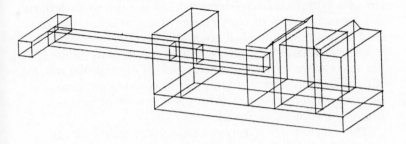

Figure 2.11 *Vice and block in the closed position*

Entries for each sub-part were made into the node and edge lists. These are the basic 'models' making up the slider. The fact that these are indeed models is defined within the computer memory by means of a model array which holds pointers to where the various sub-parts begin in the node and edge lists. Additionally they hold pointers to the fact that they are all part of a larger model, namely the sliding part. In the same way, the base part is composed of four sub-parts: the base plate, the end-stop, the guide hole and the fixed jaw. Finally, in this example, the sliding and the base parts are themselves part of the complete assembly which also contains the block to be inserted between the jaws. The tree diagram for this assembly is shown in Figure 2.12.

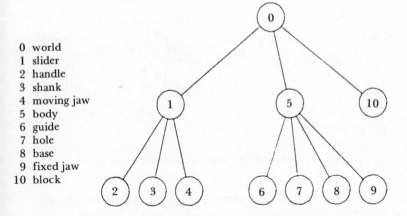

0 world
1 slider
2 handle
3 shank
4 moving jaw
5 body
6 guide
7 hole
8 base
9 fixed jaw
10 block

Figure 2.12 *Hierarchy of assembly for vice and block*

51

Now each branch of the tree corresponds to a transformation matrix. Each basic model in the system is defined in terms of its own suitably chosen frame of reference. To find its position in the full assembly, we need to transform these coordinates by each of the transformations on the branches as we move from the model up through the tree to the full assembly itself. (After this it would be necessary to perform whatever view transformations were needed to obtain a useful image for the user.)

The advantage of this type of approach is that we can change the interrelations between certain of the sub-parts while still preserving the overall configuration. For example, if we wanted to redesign the handle, it is only necessary to do this in the model space of the handle itself; the transformations needed to implant it in the full assembly would not be changed. If we wanted to move the handle relative to the shank and the moving jaw, then it is necessary only to alter the transformation operative on the branch between the handle and the moving part; this alters the way in which the handle is inserted into the model of the moving part. If we wanted to move all the moving part (for example to open or close the vice), this can be effected by changing the transformation which inserts the moving part into the full assembly. In doing this the assembly of the three sub-parts forming the moving part is not affected. Figure 2.13 shows the vice in an open position.

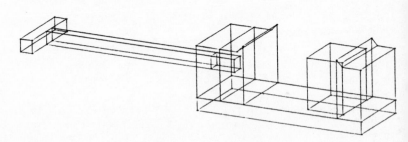

Figure 2.13 *Vice with the jaws open*

It is seen that it is important here to retain the nodes defining the basic models in their own original reference space. They should not be transformed into the space of the full assembly immediately they are input. This means that they can always be referred to by the system to obtain the most accurate definition of the shapes involved. If we transform them into global coordinates, then when the global view is changed they too will be changed and this repeated alteration will cause the build up of floating point errors due to the inherent inaccuracies of computer arithmetic. It may nonetheless be desirable to include in the node list, together with the original definition of each node, its current transformed position in global space. This saves time since the various layers of transformations do not have to be gone through repeatedly. The transformed position will of course have to be updated from the original defining position whenever one of the transformations between the model and the global space is changed.

Application to robotics

In this section we look at a means for using transformation matrices to assemble and manipulate the links of a robot. As in the last section, the essential idea is to define each link (base, shoulder, forearm, etc.) in its own reference space and then to use transformations to put them together in the correct way and to allow the display to be articulated to simulate a real robot.

The first stage is to describe a method for selecting sets of coordinate axes — each set being fixed in and moving with one of the links of the robot. While this choice can be made arbitrarily, if a consistent choice is made for each one, the resultant matrix transformations have a similar form and by changing the parameters which refer to the geometry of the particular robot used, one can arrange to simulate different makes of robot. The selection of axes used here is based upon that described in Denavit and Hartenberg (1955) and Lee (1982).

Consider a robot with n moving links; we take link 0 to be the base of the robot. We refer to the end of a link nearer the base as the 'start' and the one further away as the 'end'. Frames of reference F_i are selected ($i = 0, 1, \ldots, n$). Frame F_i moves with link i; frame F_0 is the absolute (world) frame of reference fixed in the base. We take the origin of frame F_i to be in the axis of rotation at the end of link i (which is also the start of link $i + 1$); call this origin O_i. We choose the directions of the axes as follows:

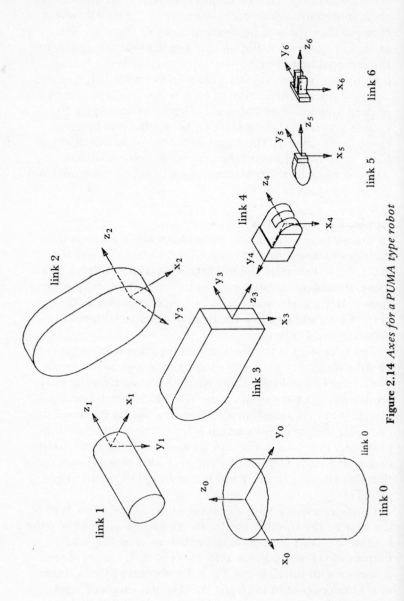

Figure 2.14 *Axes for a PUMA type robot*

- take z_i to be along the axis of rotation;
- select x_i to be perpendicular to both z_i and z_{i-1};
- select y_i to be perpendicular to both x_i and z_i to form a right handed set of axes.

Sets of axes appropriate to a PUMA type of robot are shown in Figure 2.14. Note that the selection above ensures that relative to F_i as it moves with link i, the z_{i-1} axis appears fixed.

We now look at the geometry of the robot and consider how to transform frame F_i to bring it on top of frame F_{i-1}. We do this in a three stage process:

- rotate about axis x_i to bring axis z_i parallel to axis z_{i-1}; suppose that this is through an angle $-\alpha_i$;
- rotate about the revised z_i axis to bring axis x_i parallel to axis x_{i-1} — and hence also bring axis y_i parallel to axis y_{i-1}; suppose that this rotation is through an angle $-\theta_i$;
- translate origin O_i to origin O_{i-1}; suppose that this involves moves of $-a_i$ in the x_{i-1} direction, $-b_i$ in the y_{i-1} direction and $-d_i$ in the z_{i-1} direction.

Thus α_i is the angle between z_{i-1} and z_i (in that order) as viewed along x_i. It is a constant which depends upon the geometry of the link. Angle θ_i is variable and is a measure of the rotation of link i about its 'start' axis z_{i-1}. Lengths a_i, b_i and d_i are fixed and again depend upon the geometry of the link.

The parameters introduced above can be combined into a matrix to represent a transformation of axes between frames F_i and F_{i-1}. This 4×4 matrix P_i is given below:

$$\begin{bmatrix} \cos(\theta_i) & -\cos(\alpha_i)\sin(\theta_i) & \sin(\alpha_i)\sin(\theta_i) & a_i\cos(\theta_i) - b_i\sin(\theta_i) \\ \sin(\theta_i) & \cos(\alpha_i)\cos(\theta_i) & -\sin(\alpha_i)\cos(\theta_i) & a_i\sin(\theta_i) + b_i\cos(\theta_i) \\ 0 & \sin(\alpha_i) & \cos(\alpha_i) & d_i \\ 0 & 0 & 0 & 1 \end{bmatrix}$$

The tree structure which describes the assembly of the links is very simple in this case; it is shown in Figure 2.15 for a robot with five links. The five branches in this graph each correspond to matrices of the above type. Suppose a point in space has coordinates (u, v, w) with respect to the gripper; for example, the origin of the gripper itself would have coordinates (0, 0, 0). Then the position of this point in absolute coordinates (relative to the base) is given by the product:

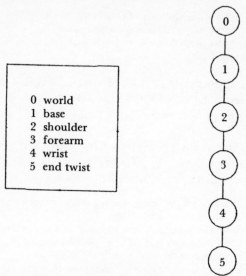

0 world
1 base
2 shoulder
3 forearm
4 wrist
5 end twist

Figure 2.15 *Hierarchy of assembly for a simple robot*

$$P_1 P_2 P_3 P_4 P_5 \begin{bmatrix} u \\ v \\ w \\ 1 \end{bmatrix}$$

In order to display the links themselves, we need of course to input their geometry into the node and edge lists. This has to be referred to the appropriate frame of reference F_i. To obtain the actual position in space, the position of each node of link i is transformed by the matrices on the branches of the tree between the link and the base, that is by the product $P_1, P_2, \ldots, P_i$. If one of the variable angles θ_j is changed, then links j, j + 1, ... are all moved but earlier ones in the tree are not. This is precisely the same effect as moving joint number j of a real robot.

This way of holding the interrelationship of the various frames of reference is very economical. There are five parameters needed to position one frame relative to the next. Additional to this of course are the data needed to describe the precise geometry of each link so that it can be drawn out. A small program for a microcomputer was developed to demonstrate some of the ideas discussed in this section. Figure 2.16 shows four plots obtained from this program for a simple straight line image of a five link robot. Between each plot just one of the link angles θ_i for some i has been changed. The robot remains intact, the links with numbers larger than i move with link i.

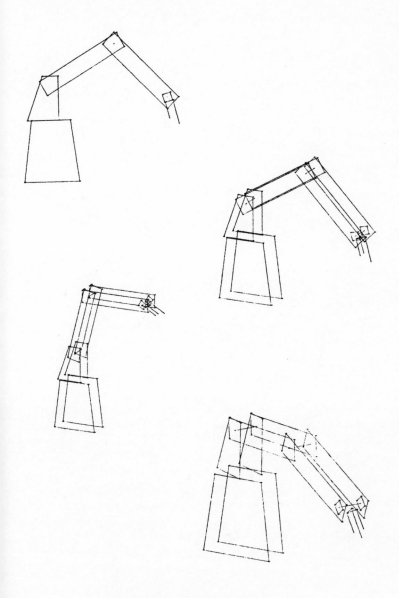

Figure 2.16 *Views of a simple robot*

CHAPTER 3

Representation of curves

Introduction
In many engineering applications, artefacts are composed
of shapes bounded by straight line segments and circular
arcs. We looked at examples with straight line segments in
Chapter 2. Shapes with lines and circular arcs are familiar and
easily defined. They are straightforward to produce manually
by turning, by numerically controlled (NC) milling machines
and so on.

For other applications however we need more complex
curves and/or surface shapes. Examples of components
requiring such entities are aerofoil sections, car body shells,
aircraft fuselages, bottles and ship hulls. In some cases curves
are needed for technological reasons, in other cases simply
because of aesthetic considerations.

There is thus a need to provide CAD systems with the
means to handle free-form curves and surfaces. It is curves
that we consider in this chapter. The techniques used to
define and manipulate them need to be simple from a
computational and mathematical point of view; easy for
a designer to use without any detailed knowledge of the
internal workings of the system; and they must allow a
reasonable variety of curve shapes to be produced. Computational simplicity can involve several factors, some of which
can be mutually incompatible, but which might include:

- ease of plotting (or cutting as part of the manufacture operation)
  the curve; usually this would be done incrementally, moving step
  by step along the curve;
- ease of obtaining intersections of curves;
- ease of determining whether a given point lies on a given curve;
- economy of the amount of data required to be stored to describe
  a curve; thus we are looking for generic curve forms.

We present ways for dealing with curve shapes on a more or less historical basis, beginning with the classical Cartesian geometry approach to the study of curves and moving on to the more recent use of parametric forms such as Bézier and B-spline curves.

Implicit equations

From a Cartesian coordinate geometry viewpoint, we would describe a circle by an equation of the form:

$$x^2 + y^2 = a^2$$

and a parabola by one of the form:

$$y^2 = 4ax$$

These are both planar curves. For a curve in three dimensions we need two equations. Thus for a helix, for example, we might have:

$$x^2 + y^2 = a^2 \qquad \text{and} \qquad y = x \tan(z/b)$$

and for a straight line the following pair of linear equations:

$$x + y + 2z = 1 \qquad \text{and} \qquad x + 2y - z = 3$$

There are several problems with this type of representation. These equations tend to be difficult to plot on normal graphical output devices. Unless the curve is very special and the device has some firmware facility for producing it directly (for example some graphics terminals can produce circular arcs for themselves), it is necessary to break the curve down into a sequence of short straight line segments and to plot these individually. Consider the example of the circle above. Given a value of x (say), we need to find the corresponding y-value in order to plot a point on the curve. In this case:

$$y = \pm (a^2 - x^2)^{1/2}$$

Our first problem is to decide which sign we need to take for the square root. Next if we want to produce incremental points along the curve, it is tempting merely to let x vary by small steps, find the corresponding y-values and plot out the results. However as Figure 3.1 shows this does not lead to even approximately equal steps along the curve itself. If we reduce the steps in x in order to bring the larger curve steps down in size, then we introduce far smaller steps than we

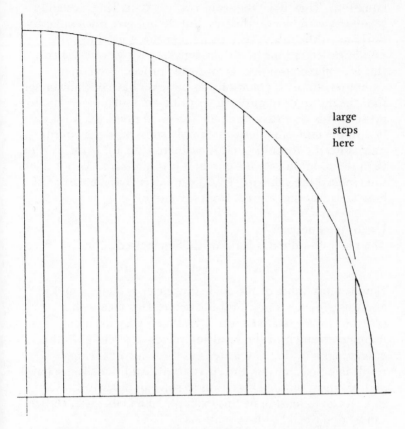

large
steps
here

Figure 3.1 *Unequal steps on subdivision along the x-axis*

need in other parts of the curve with a consequent waste of effort.

Another objection to the use of this type of equation is that it provides for no great variety of curve shape within the equations used. It is true that we could add in various extra coefficients to increase the number of parameters at our disposal. For example, the general form of a plane conic section curve is:

$$ax^2 + 2bxy + cy^2 + dx + ey + f = 0$$

However in doing this we increase the complexity of the problem of determining y for a given value of x.

The equations given in the examples here are all implicit

61

equations. They have the form $f(x, y) = 0$. They certainly relate the x and y coordinates, but do not give one explicitly in terms of the other. As a consequence we need to solve non-linear equations to obtain x given y (or vice versa) and this is a time-consuming, error-prone process even on a computer. Implicit equations do however have one advantage that the methods of curve description we look at later do not possess. It is very easy to check if a given point (X, Y) say lies on the curve in question. Simply substitute the coordinates into the formula; that is evaluate $f(X, Y)$. If this is zero, then the point lies on the curve; if it is non-zero, then it does not. Indeed the value of $f(X, Y)$ can be used as a measure of how far away the point is from the curve.

Use of parameters
The polar coordinates form of the circle used above is:

$$x = a \cos(\theta) \qquad \text{and} \qquad y = a \sin(\theta)$$

Here for any value of the angle θ, we can evaluate x and y and this provides a point on the curve. This means that plotting is comparatively easy. Just let θ vary in equal increments and plot the resulting values of x and y. In this case it happens that equal steps in the value of the angle result in equal steps on the curve itself. Additionally we have no non-linear implicit equation to solve in order to find x or y. As an extension of this example, the helix given earlier can be described by the equations:

$$x = a \cos(\theta), \qquad y = a \sin(\theta), \qquad z = b\theta$$

In these equations, the angle θ is acting as a parameter. As its value varies, the x and y (and z) values change in a consistent way and a curve is traced out. There is a problem still associated with the above parametric forms. That is that they involve trigonometric functions which take time to evaluate and can slow down the operation of the computer system. We find an indication of the way to proceed from the classical Cartesian geometry parametrization for the parabola example given above. This is:

$$x = at^2 \qquad \text{and} \qquad y = 2at$$

Here the parameter is t and both x and y are expressed as simply evaluated functions of t. An extension of this takes us to the idea of writing the coordinates as polynomials in a parameter t. Often cubic polynomials are used and these are referred to as Ferguson cubics and were introduced in 1963 (Ferguson, 1964). Usually only a segment or a curve is required and for this reason the values over which the parameter t ranges are restricted. For convenience this range is taken as being the interval from 0 to 1. (By a suitable change of variable we can always ensure that the parameter range for any curve segment is this interval.) When cubics are used the formulae for the coordinates x, y and z have the form:

$$x = a_1 + b_1 t + c_1 t^2 + d_1 t^3$$
$$y = a_2 + b_2 t + c_2 t^2 + d_2 t^3$$
$$z = a_3 + b_3 t + c_3 t^2 + d_3 t^3$$

We can write these relations more concisely by the use of vector notation and obtain:

$$\mathbf{r} = \mathbf{r}(t) = \begin{bmatrix} x \\ y \\ z \end{bmatrix} = \mathbf{a} + \mathbf{b}t + \mathbf{c}t^2 + \mathbf{d}t^3 \qquad (0 < t < 1)$$

Here **a, b, c** and **d** are coefficient vectors. Between them they have twelve components. These ensure that a great variety of curve shapes is possible. If the curve is to be planar, then we only deal with x and y and the coefficient vectors each have two components. It is this form that we use for example purposes such as the following. The parametric equation for the example is:

$$\mathbf{r} = \begin{bmatrix} x \\ y \end{bmatrix} = \begin{bmatrix} 0 \\ 1 \end{bmatrix} + \begin{bmatrix} 1 \\ 2 \end{bmatrix} t + \begin{bmatrix} 1 \\ -1 \end{bmatrix} t^2 + \begin{bmatrix} 2 \\ -2 \end{bmatrix} t^3$$

This curve segment is shown in Figure 3.2.

Parametric curves of the above type still leave some difficulties in their operation. For example, if we are given a curve or if we know approximately what a segment should look like, how do we go about choosing the various coefficient vectors? At present these vectors have no geometric meaning apart from being involved in an equation which defines the geometry of a curve. In the next section we look at a rearrangement of the Ferguson form above which allows some meaning to be attached to the vector coefficients.

The use of parametric curves introduces the problem of

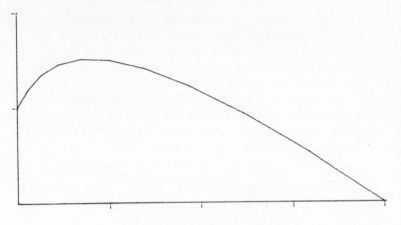

Figure 3.2 *A cubic curve segment*

determining whether a given point lies on the curve or if not how far away it is. To check if a point is on the curve it is necessary to find the corresponding value of the parameter, use this to generate the coordinates of a point on the curve and then see if these are the same as the given point. To find the parameter value we could set up three cubic equations by setting the formulae for each coordinate equal to the value of the coordinate of the given point. We thus have three simultaneous non-linear equations in a single unknown. In general these will have no common solution and we are reduced to trying to find some suitable 'best fit' solution to use. This is one of the penalties of using parametric curves; it has been found that the advantages they provide in terms of flexibility of definition, manipulation and ease of plotting outweigh such difficulties.

The Bézier formulation
The idea behind Bézier cubic curves is a simple and yet ingenious rearrangement of the Ferguson cubics discussed in the last section. Cubic polynomials are normally written in terms of t^3, t^2, t and 1. This happens to be the simplest way to express them. However we could also write a cubic in terms of the following independent functions:

$$(1-t)^3 \qquad 3t(1-t)^2 \qquad 3t^2(1-t) \qquad t^3 \qquad (3.1)$$

These are the Bézier basis functions. (In vector space terminology, they do indeed form a basis for the vector space of polynomials of degree at most three.) Each of these ones is a cubic polynomial and so is any linear combination of them; conversely any cubic can be written as a linear combination of these particular ones. Using vector coefficients as before and restricting the parameter t to lie between 0 and 1, we consider the following parametric equation for a curve segment:

$$\mathbf{r}(t) = \mathbf{a}(1-t)^3 + 3\mathbf{b}t(1-t)^2 + 3\mathbf{c}t^2(1-t) + \mathbf{d}t^3 \quad (0 < t < 1) \quad (3.2)$$

Why have we performed this rearrangement which at first sight only serves to complicate the formulae? It is because we can now attach meaning to the various coefficients involved. Consider first what happens at the start of the segment when t = 0. Here all the terms involving t as a factor vanish and we are left with:

$$\mathbf{r}(0) = \mathbf{a}$$

In the same way, at the end of the segment when t = 1 all the terms involving $(1-t)$ are zero, so:

$$\mathbf{r}(1) = \mathbf{d}$$

Thus two of the vector coefficients **a** and **d** represent end-points of the curve segment. To find out about the other two coefficients we need to differentiate Equation (3.1) with respect to t. This gives:

$$\mathbf{r}'(t) = 3[(\mathbf{b}-\mathbf{a})(1-t)^2 + 2(\mathbf{c}-\mathbf{b})t(1-t) + (\mathbf{d}-\mathbf{c})t^2] \quad (3.3)$$

Again we look at what happens at the end-points and find that

$$\mathbf{r}'(0) = 3(\mathbf{b}-\mathbf{a})$$
$$\mathbf{r}'(1) = 3(\mathbf{d}-\mathbf{c})$$

Now in general a Taylor series expansion about the value t = u for small values of h shows that:

$$\mathbf{r}(u+h) = \mathbf{r}(u) + h\mathbf{r}'(u)$$

This means that the curve leaves the point $\mathbf{r}(u)$ in the direction of vector $\mathbf{r}'(u)$; that is $\mathbf{r}'(u)$ is the tangent direction at the point where t = u. So the initial tangent direction is that of the vector $3(\mathbf{b}-\mathbf{a})$. As the scalar multiple 3 does not affect the direction of a vector, this is the direction of the vector $(\mathbf{b}-\mathbf{a})$. Now the segment is known to start at **a** and so it must head initially towards the point **b**. In the same way the tangent direction at the end of the segment is parallel to

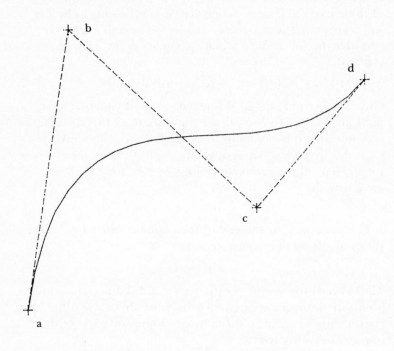

Figure 3.3 *Control points for a Bézier cubic segment*

vector $(\mathbf{d} - \mathbf{c})$, and as the end of the segment is at the point $\mathbf{d}$, we deduce that the curve must finally head towards $\mathbf{d}$ as though it were coming from the point $\mathbf{c}$. Thus we have found meaning for the other two vector coefficients $\mathbf{b}$ and $\mathbf{c}$; they lie on the tangents to the segment at its beginning and end. These ideas are summarized in Figure 3.3. The vector coefficients $\mathbf{a}$, $\mathbf{b}$, $\mathbf{c}$ and $\mathbf{d}$ are referred to as control points for the segment.

While $\mathbf{b}$ and $\mathbf{c}$ certainly lie on the end-tangents the precise distance along is not immediately obvious from looking at the shape of the segment itself. If either is moved along the corresponding tangent, the shape of the segment changes. The further away the tangent control point is from the end of the segment, the more the segment holds to the tangent direction before moving away from it. Figure 3.4 shows an

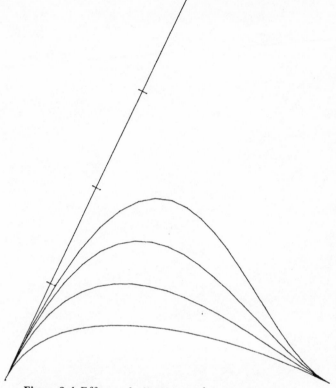

Figure 3.4 *Effects of adjusting a Bézier control point*

example of this; the three segments shown are derived from four control points, three of which are the same in each case, while the fourth, a tangent control point, is moved to three different positions along its tangent direction.

Another example is shown in Figure 3.5. Here the control points lie on the sides of a square with the end-points of the segment at opposite corners. The tangent control vectors lie along two sides and are shown in various positions together with the segments they generate. It is in this sort of configuration that the best Bézier cubic segment approximation to a quadrant of a circle is obtained and this is shown separately in Figure 3.6. The equation of this segment is:

$$\mathbf{r}(t) = \begin{bmatrix} 1 \\ 0 \end{bmatrix}(1-t)^3 + 3\begin{bmatrix} 1 \\ \alpha \end{bmatrix} t(1-t)^2 + 3\begin{bmatrix} \alpha \\ 1 \end{bmatrix} t^2(1-t) + \begin{bmatrix} 0 \\ 1 \end{bmatrix}t^3$$

where $\alpha = 4(\sqrt{2}-1)/3 = 0.552$.

67

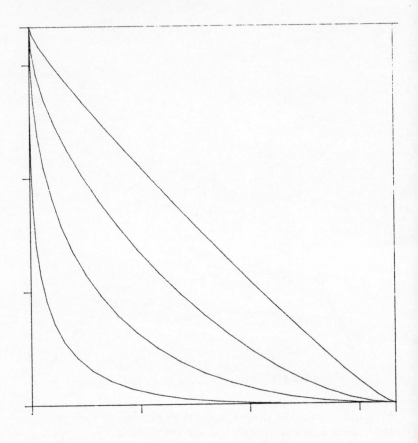

Figure 3.5 *Effects of different Bézier control point positions*

Note that this approximation to a circular quadrant is good
but is not exact. If we wanted to use Bézier segments for this
type of approximation we would need to split the curve to be
approximated into parts and fit cubic segments to each
trying at the same time to ensure that they fitted together
smoothly. This aspect is looked at in greater detail later in
this chapter.

Finally, in this section we mention the so-called convex
hull property. For a number of points in a plane, their
convex hull is the smallest convex polygon that contains
them all. Thus, for three points it is simply the triangle that

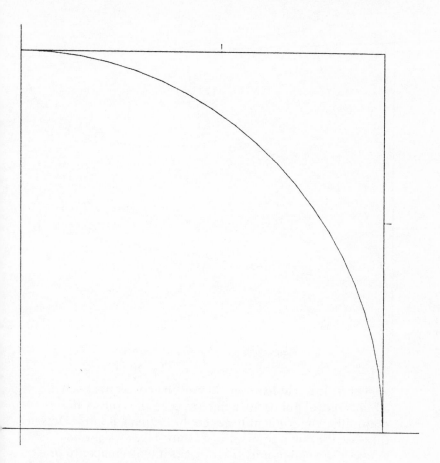

Figure 3.6 *Best Bézier cubic approximation to a quadrant of a circle*

they form; and for four points it is the quadrilateral they form unless one of them lies within the triangle formed by the other three in which case it is this triangle. The convex hull of a number of points in three-dimensional space is the smallest convex polyhedron that contains them all. For four points, this is the tetrahedron that they form.

The convex hull property for Bézier cubic curve segments says that the segment always lies within the convex hull of the four control points; this is illustrated in Figure 3.7. This applies both when the curve is planar and when it is a space curve. This property is often useful when trying to sketch

69

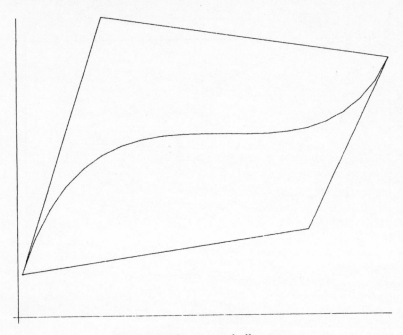

Figure 3.7 *The convex hull property*

segments. It has at least one interesting consequence. When
all the control points are collinear, then the convex hull is
simply the line segment between the extreme points. Thus in
this case the curve segment itself must be a straight line
between the end-points. It is not our intention here to prove
the convex hull property. We merely remark in passing that it
depends on a property of the choice of the basis functions
used [cf. Equation (3.1)]. This property is that their sum is
unit. To see this consider the binomial expansion of $(p + q)^3$:

$$(p + q)^3 = p^3 + 3qp^2 + 3q^2p + q^3$$

Hence the sum of the basis functions is given as follows:

$$(1 - t)^3 + 3t(1 - t)^2 + 3t^2(1 - t) + t^3 = [(1 - t) + t]^3$$
$$= 1$$

It is significant that this implies that each point on the
segment is a weighted average of the control points. That is it
has the form:

$$\lambda_1 a + \lambda_2 b + \lambda_3 c + \lambda_4 d$$

where the sum of the λ_i is unity. If these scalars were all equal (to 1/4), then the sum would be an ordinary average of the vectors. In the more general case in which they are all non-negative and sum to unity the result is a weighted average.

More general Bézier forms

The Bézier cubic segment discussed above is in fact a particular example of a whole class of Bézier forms. It happens that the cubics are most often used since they combine simplicity (in their low degree) with sufficient flexibility for useful design work. In particular the cubics do permit curve segments with points of inflexion which ones of lower degree do not. The more general Bézier segment is one of degree n; that is each of the coordinates is expressed as a polynomial of degree n in the parameter t. It requires $n + 1$ control points to be defined. These are combined to obtain the following segment equation:

$$r(t) = \sum_{i=0}^{n} \binom{n}{i} a_i t^i (1-t)^{n-i} \tag{3.4}$$

where $\binom{n}{i}$ is the binomial coefficient $(n!/[i!(n-i)!])$.

It is straightforward to check that when $n = 3$ this form reduces to that of Equation (3.2). It should be noted that Equation (3.4) requires that zero raised to the power zero be treated as unity.

A number of properties of the cubic segment carry over to this more general case. For example, the values of r and its first derivative are as follows:

$$r(0) = a_0 \qquad\qquad r(1) = a_n$$
$$r'(0) = n(a_1 - a_0) \qquad r'(1) = n(a_n - a_{n-1})$$

Thus the segment begins at a_0 and ends at a_n, that is at the first and last of the control points. The initial tangent direction is from the start point towards a_1, and the final tangent direction is from a_{n-1} to the end-point.

It is interesting to note that the form of the first derivative of the segment of degree n can itself be written in Bézier form but with degree $n - 1$ and with control vectors which are differences of the original ones. The derivative of Equation (3.4) is:

$$r'(t) = \sum_{i=0}^{n-i} \binom{n-1}{i} (a_{i+1} - a_i) t^i (1-t)^{n-i} \qquad (3.5)$$

We could form the second derivative in the same sort of way; the typical control vector in this case would be:

$$a_{i+2} - 2a_{i+1} + a_i$$

It is interesting to note that at each stage of differentiation, the new control vectors are finite difference combinations of the original ones. As the second derivatives of the position vector r determine the curvature, we note the values of this derivative at the ends of the segment:

$$r''(0) = a_2 - 2a_1 + a_0 \qquad (3.6)$$
$$r''(1) = a_{n-2} - 2a_{n-1} + a_n$$

The basis functions used in Equation (3.4) are the following:

$$\binom{n}{i} t^i (1-t)^{n-i} \qquad i = 0, 1, \ldots, n$$

These are the terms in the formal binomial expansion of $[(1-t) + t]^n$. As a result the sum of the basis functions is unity and the functions themselves are all non-negative (when t lies between 0 and 1). Thus every point on the segment is a weighted average of the control vectors and as a result the curve lies entirely within the convex hull of the control vectors.

An example of a seventh degree Bézier curve segment is given in Figure 3.8. The effects of moving one of the control points are shown; note that the whole of the segment is affected by this change.

Manipulating Bézier segments

The Bézier form is an efficient way of storing curve information. We concentrate here upon the cubic form, although the ideas go across to the more general case. A cubic segment is defined by four control points and these can be inserted into the node list defining the geometry of the shapes being created in the CAD system. We need to identify that all four are related to the same curve entity, so a structure more subtle than the simple edge list introduced in Chapter 2 is

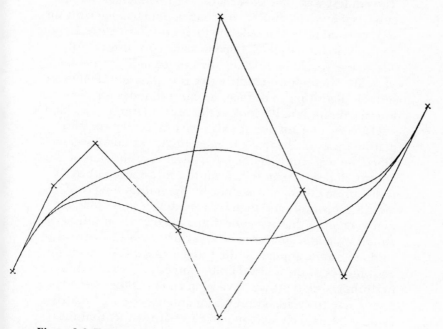

Figure 3.8 *Two seventh degree Bézier segments and control polygons*

required. Perhaps a structure could be used in which the various entities (line segments, curve segments, etc.) that are used are listed with a code to describe which type each is and a pointer to a separate list of node numbers which are those involved with the entity. Thus a line segment would imply that two nodes from this list are to be taken, and a cubic segment would need four. An example of this type of structure is shown below.

Node list			*Entity list*		*Nodes involved in entities*
x	y	z	Code	pointer	Node number
1	1	0	1	1	1
4	1	0	1	9	2
4	3	0	2	5	6
3	6	0	1	3	1
2	6	0			3
1	3	0			4
					5
					6
					2
					3

The crudest way that allows a user to manipulate a Bézier cubic is merely to display the control points along with the segment and have him redefine their positions. Ultimately the controls do have to be defined in some sort of way. An alternative would be to allow the user to indicate a number of points through which the curve is to pass and then use a suitable algorithm to deduce the control nodes for the internal definition. We look at this aspect later in this chapter.

Once the control points exist, we can use the transformation ideas of Chapter 2 to manipulate the curve segment. Since the segment is defined as a linear combination of the control nodes, we can transform the entire segment by transforming these nodes alone by pre-multiplying by a matrix transform (and then recalculating points along the new version so that it can be displayed on the output device). Rotations and scalings can be described by 3×3 matrices, so these can be applied to the three-dimensional Cartesian coordinates of the control points directly. If we need to perform translations we can add in an extra homogeneous coordinate to each control vector and use the 4×4 matrix form. Alternatively we can make use of the fact that the Bézier segment is a weighted average of the control points. The coefficient polynomials add to unity; so adding the same vector to each control node has the same effect as adding that vector onto the whole sum. Thus we can translate the whole curve merely by translating each control node by the appropriate amount.

Bézier segments with homogeneous coordinates
When we discussed transformations in Chapter 2, we looked at the advantages of using homogeneous coordinates to allow a unification of the various types of transformation in matrix terms. Their use also provides some more flexibility in the types of curve segment that can be defined. It is this that we consider next.

An ordinary Bézier cubic segment as in Equation (3.2) is defined by four control points, each determined by three Cartesian coordinates. The x, y, z coordinates of a point on the curve are a combination of the relevant components. Suppose instead that the control points are specified in terms of homogeneous coordinates so that each control vector on the right hand side of Equation (3.2) has four components.

That equation tells us how to combine these for each value of the parameter t to produce a vector with four components X, Y, Z and W. By forming the quotients X/W, Y/W and Z/W we obtain the Cartesian point in space represented by the homogeneous coordinates. As t varies, so the point moves and traces out a curve segment as before.

The simplest way to convert an ordinary Bézier segment into one with homogeneous coordinates is to add in a fourth component of unity to each of the control vectors. Thus these represent the same Cartesian points as before. When we form the X, Y, Z and W components of a point on the curve using the equivalent of Equation (3.2), we find that W is one for any value of the parameter t; this is because the sum of the Bézier basis functions [Equation (3.1)] is unity. Thus the point represented on the segment when we divide by W is precisely the same as with the original ordinary Bézier segment, and we have generated the same curve shape.

However, if we take the new homogeneous form of any control point obtained by adding in the extra component of one, we can multiply each component by a scalar and not change the Cartesian point it represents. We can do this with each control point and we need not use the same scalar for each one. When we combine these as in Equation (3.2), W is no longer constant as t varies, and we obtain a different curve shape. By changing the scalar used to modify each control vector, we can alter the curve shape without changing the Cartesian position of the control points.

As we discuss later, for a segment in homogeneous form, the Cartesian positions of the control points still govern the curve shape in the same sort of way as before; notably they determine the end-points and the end-tangent directions. Changing the homogeneous representation of the control points (without moving the points they represent) allows us to vary the curve shape while preserving the end-conditions. This is where the extra flexibility comes from. It is possible in some situations to use homogeneous forms of lower degree and still provide an adequate supply of curve shapes. In particular, we can represent exactly segments of conic section curves, including circular arcs, by homogeneous Bézier segments of degree two. The extra freedom gained with using homogeneous coordinates means that more unknowns have to be taken into account. Perhaps it is for this reason that not many commercial CAD systems seem (overtly at least) to allow the user access to homogeneous Bézier segments.

It is noted previously that if the extra homogeneous coordinate of each control vector is unity, then the curve is essentially the same as an ordinary Bézier segment. We now try to illustrate the above ideas by looking at the effects of using control vectors where the homogeneous coordinate is not one. For convenience, we retain unity as the fourth coordinate for the vectors at the ends of the segment; this can help in the task of putting segments together as it ensures that the actual Cartesian end-points of segments have positions specified by the first three coordinates.

The example considered is based upon a circular arc. The idea of approximating a circular quadrant by a Bézier cubic is considered earlier; it cannot be done exactly. Now consider the approximation by a Bézier quadratic segment. The smaller degree means that using the ordinary form cannot produce better results. Suppose the arc has its centre at the origin $(0, 0)$ and goes from the point $(1, 0)$ to $(0, 1)$. As there are only three control vectors for a quadratic segment, the middle one has to provide the tangent direction at both ends of the segment. So it must lie on the intersection of the end-tangents and in our example this is at $(1, 1)$. So the ordinary Bézier control vectors are selected already. It is written below in homogeneous form with all the homogeneous coordinates set to one; because we are dealing with a planar curve, the z-coordinate is omitted and the homogeneous coordinate is the third one. Note also that the value of W shown is also unity as indicated above:

$$\mathbf{R}(t) = \begin{bmatrix} X \\ Y \\ W \end{bmatrix} = \begin{bmatrix} 1 \\ 0 \\ 1 \end{bmatrix}(1-t)^2 + 2\begin{bmatrix} 1 \\ 1 \\ 1 \end{bmatrix}t(1-t) + \begin{bmatrix} 0 \\ 1 \\ 1 \end{bmatrix}t^2$$

which implies that:

$$X = 1 - t^2 \qquad Y = 2t - t^2 \qquad W = 1$$

This segment is shown in Figure 3.9 and it is not precisely circular. We now look at the effect of changing the middle control vector to $[\alpha, \alpha, \alpha]^T$ for some suitable scalar α. The Cartesian point represented by this vector is $(1, 1)$, so it still lies on the intersection of the tangents. With a little experimentation we find that if we take α to be $1/\sqrt{2}$ we obtain

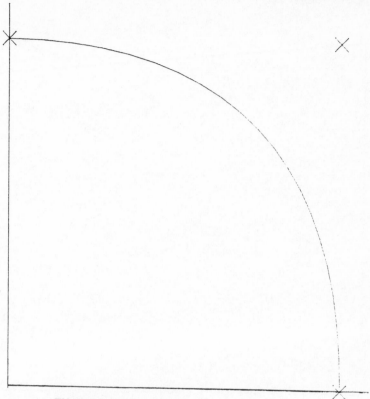

Figure 3.9 *Approximation to a circular quadrant*

the segment shown in Figure 3.10. The equation for it is as follows:

$$\mathbf{R}(t) = \begin{bmatrix} X \\ Y \\ W \end{bmatrix} = \begin{bmatrix} 1 \\ 0 \\ 1 \end{bmatrix}(1-t)^2 + 2\begin{bmatrix} 1/\sqrt{2} \\ 1/\sqrt{2} \\ 1/\sqrt{2} \end{bmatrix}t(1-t) + \begin{bmatrix} 0 \\ 1 \\ 1 \end{bmatrix}t^2$$

which implies that:

$$X = 1 - \sqrt{2}(\sqrt{2}-1)t - (\sqrt{2}-1)t^2$$
$$Y = \sqrt{2}t - (\sqrt{2}-1)t^2$$
$$W = 1 - \sqrt{2}(\sqrt{2}-1)t - \sqrt{2}(\sqrt{2}-1)t^2$$

The reader can check that $X^2 + Y^2 = W^2$ and this shows that when we convert the points on the segment to ordinary Cartesian points $(X/W, Y/W)$ they do indeed lie on the arc of a circle.

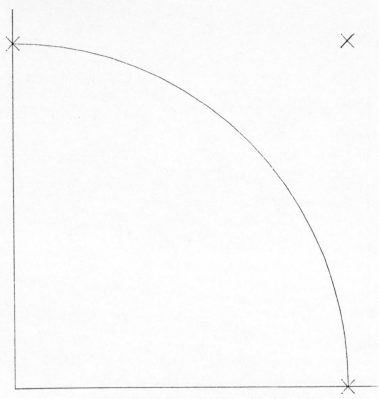

Figure 3.10 *Exact circular quadrant*

If we apply scaling in the x-direction, say doubling all the x-coordinates, then the resulting control vectors are:

$$\begin{bmatrix} 2 \\ 0 \\ 1 \end{bmatrix} \begin{bmatrix} \sqrt{2} \\ 1/\sqrt{2} \\ 1/\sqrt{2} \end{bmatrix} \begin{bmatrix} 0 \\ 1 \\ 1 \end{bmatrix}$$

and these generate a quadrant of an ellipse. In fact any conic section segment can be represented by a Bézier quadratic of the above form. Each of the Cartesian coordinates (x, y) of points on the curve is a quotient of two polynomials in t, that is a rational function of t. Thus the segment is in the rational Bézier form.

The same properties apply to the rational form as to the ordinary one. The ends of the segment are the Cartesian points determined by the end control vectors, and the end-tangent directions are those determined by the Cartesian points corresponding to the pair of control vectors at either end of the range of control vectors. To check that these results do indeed hold it is necessary to use differentiation; on this occasion the appropriate derivatives of quotients need to be found. The convex hull property also applies: the rational Bézier segment lies within the hull of the Cartesian points determined by the control vectors. In particular for rational quadratics, there are only three control vectors and so the corresponding hull is a triangle which is a plane figure. So the segment produced is necessarily a planar curve lying in the plane of the points specified by the control vectors. This remark of course applies also to the ordinary Bézier segments and this is one of the reasons why Bézier cubics are so popular; they are the lowest degree curve which permits genuine three-dimensional curves to be generated.

Finally in this section, we look again at circular arcs. If we wish to represent a given arc in rational quadratic form, then it is straightforward to select the control vectors. The ends of the arc provide two of these and we can take the homogeneous coordinate to be unity. The intersection of the end-tangents provides the starting point for selecting the third. We need also to select its homogeneous coordinate. If 2ϕ is the angle subtended at the centre by the radii from the two end-points, then the required value is $\cos(\phi)$; this value should be multiplied onto the values of the coordinates of the tangent intersection point and added in as the last coordinate.

There is an interesting case when we are dealing with a semicircle. Here the value of ϕ is $90°$ and we have zero as the homogeneous coordinate. The geometry of the situation is shown in Figure 3.11; here it is seen that the end-tangents are parallel and do not intersect, at least not in the finite part of the plane. They can be regarded as intersecting at infinity and this is what the homogeneous coordinate of zero corresponds to. The segment can still be represented in rational Bézier form but a little more care is required in this case in the choice of the middle control vector. By symmetry considerations we need to take a point on the line midway between the parallel end-tangents. Thus in the case of the example in Figure 3.11 this gives a homogeneous vector of

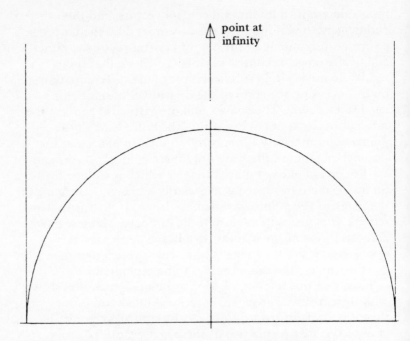

Figure 3.11 *Control points for a semicircle*

the form $[0, a, 0]^T$ for some suitable value of a. By evaluating the point on the segment corresponding to $t = 0.5$ it is easy to check that we need to take $a = 1$ here. This gives the following segment equation which exactly reproduces the semicircular arc shown:

$$\mathbf{R}(t) = \begin{bmatrix} 1 \\ 0 \\ 1 \end{bmatrix} (1-t)^2 + 2 \begin{bmatrix} 0 \\ 1 \\ 0 \end{bmatrix} t(1-t) + \begin{bmatrix} -1 \\ 0 \\ 1 \end{bmatrix} t^2$$

The de Casteljau algorithm
There is a recursion formula which relates the Bézier basis functions of differing degrees. If we use the following notation to denote these polynomials functions:

$$b_{ni}(t) = \binom{n}{i} t^i (1-t)^{n-i}$$

then the relationship is:

$$b_{ni}(t) = t b_{n-1,i-1}(t) + (1-t) b_{n-1,i}(t)$$

The de Casteljau algorithm (de Casteljau, 1985; Baer *et al.*, 1979) makes use of this result to provide a way for obtaining points on Bézier segments by repeated linear interpolation. It works for any degree curve and for rational forms as well; we show it here for an ordinary Bézier cubic segment. Suppose the control points are denoted by a_0, b_0, c_0 and d_0. For each value of t, we develop a triangular array of vectors as shown below:

$$
\begin{array}{ccccccc}
a_0 & & & & & & \\
 & a_1 & & & & & \\
b_0 & & a_2 & & & & \\
 & b_1 & & a_3 & & & \\
c_0 & & b_2 & & & & \\
 & c_1 & & & & & \\
d_0 & & & & & & \\
\end{array}
$$

In columns after the first, each entry is a point linearly interpolated between the two entries to its left, this interpolation being dependent on t. Thus we have for example:

$$a_1 = (1 - t)a_0 + tb_0$$
$$b_2 = (1 - t)a_1 + tc_1$$

and the other entries similarly.

If we connect together in pairs the original four control points to obtain three line segments which form what is sometimes called the Bézier polygon, then the second column represents points on these lines. Similarly if these are joined, then the third column entries represent two points on the two new lines. This is shown in Figure 3.12. The significant fact is that the last point obtained a_3 is a point on the curve segment; it is the point at parameter value t. This follows from the recursion formula for the basis functions or, equivalently, it can be checked in this case by writing down algebraic expressions for each of the vectors in the table from their definitions.

Another important result derives from the above construction. If we take a point on a Bézier curve, then it splits the segment into two pieces. Each piece has coordinates that are polynomials in the parameter t. For both subsegments, by using a simple change of variable, the parameter can be made to run over the values between 0 and 1 and then the subsegment can be itself expressed in Bézier form. In this way each subsegment has its own control points. When the segment is split at the point with parameter value t, these

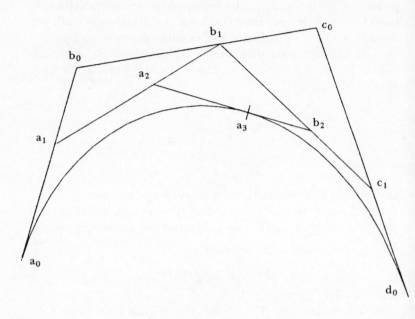

Figure 3.12 *Points for the de Casteljau algorithm*

control points turn out to be the vectors in the above triangular table running along the top and bottom sloping sides, that is the points:

$$\begin{array}{cccc} a_0 & a_1 & a_2 & a_3 \\ a_3 & b_2 & c_1 & d_0 \end{array}$$

That these are reasonable points is seen also in Figure 3.12; for instance it is clear that these points give the appropriate tangency conditions at the ends of the two subsegments.

This subdivision technique has application when we need to find intersections of curves; we see this in Chapter 4. It is also useful for displaying complete Bézier segments on graphical output devices. If we repeatedly subdivide the curve, say at the point corresponding to the mid-value of the parameter, then we obtain a sequence of sets of control

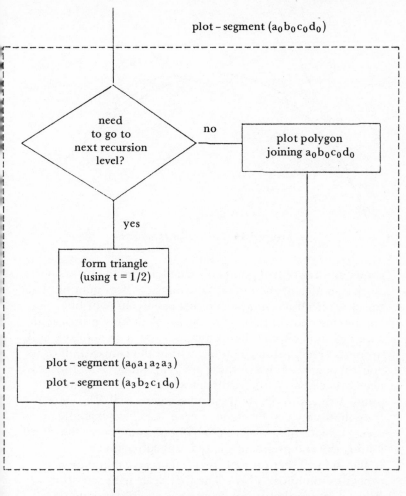

Figure 3.13 *Recursive plotting algorithm*

points. These points converge to points on the segment itself and do so quite rapidly. Subdivision at the mid-value of the parameter means that we do not have to bother with the parameter itself; the triangular array is generated merely by taking averages. It leads to a simple recursive plotting algorithm whose basic outline is shown in Figure 3.13.

We could stop the recursion when it has reached some predetermined level and at that stage the plot is produced by drawing a straight line between the control points. An example of this approach is shown in Figure 3.14; in fact this

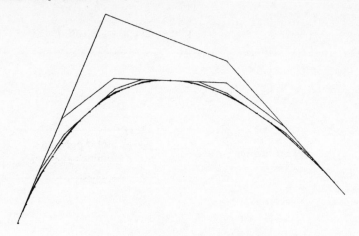

Figure 3.14 *Plotting by subdivision*

shows all the control points obtained during the recursion
to give an idea of the rate of convergence. Bear in mind that
the curve itself has not been plotted; only straight line
connecting control points are shown. A more sophisticated
approach to stopping the recursion is to use the convex hull
property. Each subsegment lies within the convex hull of its
control points. So when these are approximately collinear we
need proceed no further: the lines between the control
points will approximate their subsegment well. Thus a test
for collinearity can be made at each stage, for example by
seeing how far the tangent control points are from the chord
joining the end-points of the relevant subsegment.

We end this section with two algebraic manipulation
processes like those above. The first deals with rewriting a
Bézier form so that it is presented as one of (apparently)
higher degree. This relates to the fact that a polynomial of
any degree is certainly a polynomial of higher degree (with
zeros as its leading coefficients. If **a, b, c** and **d** are the four
control points of a Bézier cubic segment, we define new
vectors by:

$$\mathbf{a}' = \mathbf{a}$$
$$\mathbf{b}' = [\mathbf{a} + 3\mathbf{b}]/4$$
$$\mathbf{c}' = [2\mathbf{b} + 2\mathbf{c}]/4$$
$$\mathbf{d}' = [3\mathbf{c} + \mathbf{d}]/4$$
$$\mathbf{e}' = \mathbf{d}$$

If these are used as control points of a Bézier quartic, then they define the same curve shape. Moreover, they actually define precisely the same curve; for if the algebraic expression for the quartic is expanded the highest degree terms all cancel and the original cubic remains.

The procedure extends to any degree of Bézier segment. The new control points are again linear interpolations between pairs of previous ones; the factors for a curve of degree n used to find the ith new control point of the version of degree $(n + 1)$ are $i/(n + 1)$ and $(n + 1 - i)/(n + 1)$ for $i = 0, 1, \ldots, n + 1$.

An application of this procedure can be seen by considering the following situation. Suppose a cubic segment has been positioned so that it has smooth joins onto other geometric entities at its two ends. The curve however is not quite of the form required. By increasing its degree to 4 we do not change its shape, but we introduce an additional control point, the middle one. This we can move to adjust the curve shape without changing the end-points or the end-tangent directions.

On a related topic we now look at a method for writing polynomial expression in Bézier form. It follows a Pascal's triangle idea. In the usual situation, entries are created in the rows of a triangle, each one being the sum of the two immediately above it. Each row in one entry is longer than the previous one; the entry at an end is a repetition of the end entry of the previous row. If we are given a polynomial, then we start with the first row consisting of the constant term. From this the triangle is created. The difference is that at each stage the entry at the right hand end of a row is increased by the coefficient of the corresponding term in the polynomial. The procedure is probably easier to appreciate from an example, and we consider two polynomials:

$$x(t) = 2 + 3t - 2t^2 + t^3$$
$$y(t) = t - t^2$$

The triangles formed are the following:

```
            2                                   0
         2      2 + 3 = 5              0     0 + 1 = 1
       2    7     5 - 2 = 3           0   1    1 - 1 = 0
     2   9   10    3 + 1 = 4         0   1   1   0 + 0 = 0
   2  11  19  14   4 + 0 = 4        0   1   2   1   0 + 0 = 0
```

85

The entries for the Bézier control points are obtained from the row corresponding to the degree of the Bézier expression; we need to divide each by the appropriate binomial coefficient. Thus the Bézier cubic form for the segment defined by the two polynomial coordinate functions above is found from the penultimate row in each of the above triangles:

$$\begin{bmatrix} 2 \\ 0 \end{bmatrix}(1-t)^3 + 3\begin{bmatrix} 3 \\ 1/3 \end{bmatrix}t(1-t)^2 + 3\begin{bmatrix} 10/3 \\ 1/3 \end{bmatrix}t^2(1-t) + \begin{bmatrix} 4 \\ 0 \end{bmatrix}t^3$$

and, with (apparently) increased degree, the quartic form determined by the last rows of the triangles is:

$$\begin{bmatrix} 2 \\ 0 \end{bmatrix}(1-t)^4 + 4\begin{bmatrix} 11/4 \\ 1/4 \end{bmatrix}t(1-t)^2 + 6\begin{bmatrix} 19/6 \\ 1/3 \end{bmatrix}t^2(1-t)^2 + 4\begin{bmatrix} 7/2 \\ 1/4 \end{bmatrix}t^3(1-t) + \begin{bmatrix} 4 \\ 0 \end{bmatrix}t^4$$

These two expressions are of course identical as would be seen if they were expanded out. They represent an example of two seemingly different curve forms which generate the same shape. If higher degrees than cubics are used it is possible to produce forms which are algebraically distinct but which nonetheless generate the same curve shape. The algebraic form depends upon the way in which the parameter is used. It is important to bear in mind that the parameter is only a means for generating the curve; it has no geometric significance and has no effect upon the shape of the curve which is what we are mainly interested in. This means that it is theoretically possible to change the parameterization (for example by writing t as a function of another variable) and obtain a form with better properties while still retaining the shape that is defined. Some work in this area has been undertaken (Ball, 1984; Mullineux, 1982a, 1982b).

Introduction to B-spline curves

With Bézier cubics we have just four control vectors with which we can change the appearance of the segment. While this is quite sufficient for reasonably short segments without too many peculiarities, there is a need also to be able to model more sophisticated curves. From a Bézier point of view, we could join segments together and this approach we look at later in this chapter. We could also increase the degree of the curve as we saw earlier. This certainly gives more control points to play about with, but there are disadvantages in using curves of degree higher than is needed (oscillations can result) and manipulation of any control point affects the whole of the segment shape.

It is this latter point which is one of the motivations for using B-splines. This type of curve form provides a description in terms of control vectors, but moving the position of one only alters a particular portion of the shape. We thus have local control over the segment. To see why Bézier curves do not possess this property, we look again at the Bézier basis functions in Equation (3.1). These are plotted out in Figure 3.15. Each is continuous with continuous

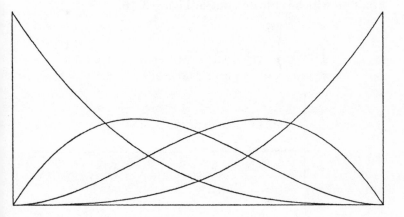

Figure 3.15 *Cubic Bézier basis functions*

derivatives, and this guarantees that the segment itself is continuous, with continuously varying curvature. However each basis function is non-zero over the whole interval from 0 to 1. This means that any change in any control vector carries through to result in a change in the segment. This remains true even if we increase the degree of the segment. We are going to make sure that the basis functions for B-splines have the continuity properties but also that they are zero over a large part of their interval of definition.

To see how to do this we first look at cubic spline functions. These are functions of a parameter t which are piecewise cubic polynomials; that is they are cubics over certain consecutive intervals but the coefficients of the cubic

87

change from one interval to the next. We need to make sure that the various cubic pieces join up in such a way that the whole function is continuous and has continuous first and second derivatives. This degree of continuity is all that is often required for design purposes. The points at which the form of the function changes are called knots.

More precisely then, a cubic spline on knots $t_0, t_1, \ldots, t_k$ is a function $f(t)$ which is continuous and has continuous first and second derivatives and is a cubic polynomial for $t_{i-1} < t < t_i$ (for $i = 1, 2, \ldots, k$).

An example on the knots 0, 1 and 3 is the following function which is plotted out in Figure 3.16:

$$f(t) = \begin{cases} 0 & t < 0 \\ t^3 & 0 < t < 1 \\ -7t^3 + 24t^2 - 24t + 8 & 1 < t < 3 \\ 3t^3 - 66t^2 + 246t - 262 & 3 < t \end{cases}$$

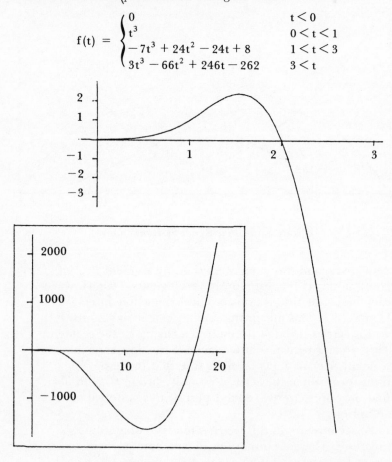

Figure 3.16 *Example of a piecewise cubic polynomial function*

It is straightforward, if time consuming, to check that the function and its first two derivatives are continuous in this case. However this is not necessary if we consider how such functions can be built up. We need the idea of Macaulay brackets (which are often used to study the bending of beams); these are used to define the following function:

$$[t-a]^n = \begin{cases} 0 & \text{if } t < a \\ (t-a)^n & \text{if } t > a \end{cases}$$

So for example, when $n = 1$, we have a ramp function which begins when $t = a$; and we include the case when $n = 0$, which corresponds to a unit step at $t = a$. By its construction, this function and its first $(n-1)$ derivatives are zero at $t = a$ (on both sides of the join). Hence the function and these derivatives are continuous everywhere and the function is a piecewise polynomial of degree n. Now by its definition, if we take a cubic spline and look at two pieces on either side of a knot, their difference is a cubic polynomial which is zero at the knot and whose first and second derivatives are also zero there. Thus this difference is precisely a scalar multiple of a cubic Macaulay bracket function, so we pass from one piece of a cubic spline to the next by adding (or subtracting) multiples of cubic Macaulay bracket functions. In this way the most general cubic spline on the knots 0, 1, 3 (which is zero for $t < 0$) is:

$$f(t) = \lambda[t-0]^3 + \mu[t-1]^3 + \nu[t-3]^3$$

In the above example, λ, μ and ν are 1, -8 and 10 respectively.

B-spline basis functions

Consider the following cubic spline function on the knots 1, 2, 3, 4, 5:

$$f(t) = \left\{ [t-1]^3 - 4[t-2]^3 + 6[t-3]^3 - 4[t-4]^3 + [t-5]^3 \right\} /6$$

This function is certainly zero for $t < 1$. What is remarkable is that we have been able to choose the coefficients so that it also vanishes for $t > 5$; to check this it is necessary to multiply out all the bracketed terms. The function is plotted out in Figure 3.17. It is a piecewise cubic which vanishes outside the interval from 1 to 5 and is non-zero over four intervals between the knots. It is found that for a cubic this is

89

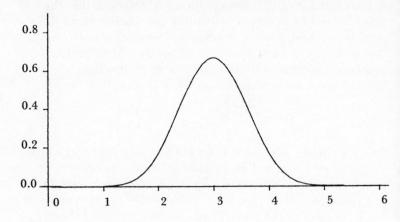

Figure 3.17 *Smooth cubic function only non-zero over a limited range*

the smallest number of intervals over which such a function
can be non-zero without being zero everywhere. The unsigned
coefficients in front of the Macaulay brackets inside the main
brackets are 1, 4, 6, 4, 1 which are the binomial coefficients
in the expansion of a quartic expression. The reason why the
whole expression is divided by 6 is that it has been normalized
so that:

$$\int_{-\infty}^{\infty} f(t)\, dt = 1$$

The fact that the coefficients in the above example come out
conveniently to binomial coefficients is in part due to the
fact that the knots are equally spaced. We can define
functions of this type for any set of knots. The important
thing to aim for is that the functions should be non-zero only
over four consecutive intervals. This type of function is the
B-spline basis function for the given set of knots. There is a
deal of computation involved in the above expression for
$f(t)$. We now discuss a computationally simpler method of
obtaining the basis functions, although at first sight it looks
enormously more complex. It should be noted that a lot of
the numbers involved in the arithmetic are in fact going to be
zero (as is seen in the later examples) and it is this that leads
to the computational simplicity. This method of obtaining
the basis functions is the Cox-de Boor algorithm. We present

it here as a means for obtaining cubic basis functions, although it will extend to other degrees. The knots used are $t_0, t_1, t_2, \ldots$ and we use $p_i(t)$ to denote the basis function that is non-zero between t_i and t_{i+4}. The algorithm determines the value of $p_i(t)$ for any given value of t. First define a_1, b_1, c_1 and d_1 as follows:

$$a_1 = \begin{cases} (t_{i+1} - t_i)^{-1} & \text{if} \quad t_i < t < t_{i+1} \\ 0 & \text{if not} \end{cases}$$

$$b_1 = \begin{cases} (t_{i+2} - t_{i+1})^{-1} & \text{if} \quad t_{i+1} < t < t_{i+2} \\ 0 & \text{if not} \end{cases}$$

$$c_1 = \begin{cases} (t_{i+3} - t_{i+2})^{-1} & \text{if} \quad t_{i+2} < t < t_{i+3} \\ 0 & \text{if not} \end{cases}$$

$$d_1 = \begin{cases} (t_{i+4} - t_{i+3})^{-1} & \text{if} \quad t_{i+3} < t < t_{i+4} \\ 0 & \text{if not} \end{cases}$$

And then evaluate the following:

$$a_2 = [(t - t_i)a_1 + (t_{i+2} - t)b_1]/(t_{i+2} - t_i)$$
$$b_2 = [(t - t_{i+1})b_1 + (t_{i+3} - t)c_1]/(t_{i+3} - t_{i+1})$$
$$c_2 = [(t - t_{i+2})c_1 + (t_{i+4} - t)d_1]/(t_{i+4} - t_{i+2})$$

$$a_3 = [(t - t_i)a_2 + (t_{i+3} - t)b_2]/(t_{i+3} - t_i)$$
$$b_3 = [(t - t_{i+1})b_2 + (t_{i+4} - t)c_2]/(t_{i+4} - t_{i+1})$$

$$a_4 = [(t - t_i)a_3 + (t_{i+4} - t)b_3]$$

The required value of $p_i(t)$ is a_4. Note that there is no division at the last stage here.

In cases where some of the knots are repeated, the algorithm occasionally demands that the quotient of zero by zero be found. The convention is that this should be treated as zero. From a computational point of view, whenever a quotient is to be found, the numerator is checked; if it is zero the result is taken to be zero irrespective of the value of the denominator.

We now present some examples of the evaluation of B-spline basis functions. It should be noted that these are not themselves the free-form curve that we are seeking to produce, they are the means to that end. In practice on a

CAD system, the calculations we are about to present are hidden away in the internal operations. We discuss these examples nonetheless as they help to provide insight into how CAD systems work and the limitations of the methods used.

When we show the intermediate values we do so in a triangular display as follows:

$$
\begin{array}{ccccccc}
a_1 & & & & \\
& a_2 & & & \\
b_1 & & a_3 & & \\
& b_2 & & a_4 \\
c_1 & & b_3 & \\
& c_2 & & \\
d_1 & & &
\end{array}
$$

In this way the calculations are revealed as being part of a process which is an extension of the way in which finite difference tables are formed. At each stage, two entries in one column are combined to give an entry in the next until only one value remains. In particular, if the two entries are both zero, then their resultant is also zero.

When dealing with B-splines we always need to select a suitable knot set. The way in which this is done is often fairly arbitrary and in commercial CAD systems which use B-splines very often the user is not allowed to have any influence over the choice. One possible selection is to have them equally spaced at unit intervals with repetitions at the ends of the sequence. Thus to obtain $(n + 1)$ basis functions we define $(n + 5)$ knots ranging over an interval of length $(n - 2)$; the knots are:

$$t_0 = 0, 0, 0, 0, 1, 2, 3, \ldots, n - 4, n - 3, n - 2, n - 2, n - 2, t_{n+4} = n - 2$$

The Cox-de Boor algorithm permits the repetition of knots. This set of knots is used in examples 2 and 3 below and provides basis functions with good properties.

Example 1 This is based upon the example introduced at the beginning of this section. The knots used are $t_1 = 1$, $t_2 = 2$, $t_3 = 3$, $t_4 = 4$ and $t_5 = 5$ and in the notation of the algorithm we are evaluating p_i at various values of t. We will use two values: $t = 2.25$ and $t = 3$.

For $t = 9/4$:

$$
\begin{array}{ccccc}
0 & & & & \\
& 3/8 & & & \\
1 & & 22/96 & & \\
& 1/8 & & 121/384 \\
0 & & 1/96 & \\
& 0 & & \\
0 & & &
\end{array}
$$

For t = 3:

$$\begin{matrix}
0 & & & \\
 & 0 & & \\
1 & & 1/6 & \\
 & 1/2 & & 2/3 \\
0 & & 1/6 & \\
 & 0 & & \\
0 & & & \\
\end{matrix}$$

Thus we have that $p_1(9/4) = 121/384$ and $p_1(3) = 2/3$. Both of these agree with the values obtained by substituting into the example given at the beginning of this section.

Example 2 Figure 3.18 shows the nine B-spline basis functions defined over the interval from 0 to 6. The 13 knots used are:

$$t_0 = 0, 0, 0, 0, 1, 2, 3, 4, 5, 6, 6, 6, t_{12} = 6$$

and these are chosen in accordance with the scheme outlined before the last example. (In that notation, we have n = 8.)

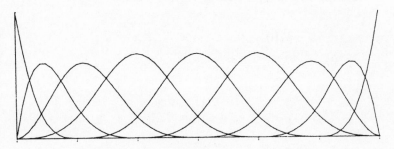

Figure 3.18 *Cubic B-spline basis functions*

Below we list the values of these functions at various points in the interval and note that the sum of the values of all the basis functions at each point is unity:

i	$p_i(0)$	$p_i(1.5)$	$p_i(3.0)$	$p_i(4.5)$	$p_i(6.0)$
0	1.000	0.000	0.000	0.000	0.000
1	0.000	0.031	0.000	0.000	0.000
2	0.000	0.469	0.000	0.000	0.000
3	0.000	0.479	0.167	0.000	0.000
4	0.000	0.021	0.667	0.021	0.000
5	0.000	0.000	0.167	0.479	0.000
6	0.000	0.000	0.000	0.469	0.000
7	0.000	0.000	0.000	0.031	0.000
8	0.000	0.000	0.000	0.000	1.000
Sum	1.00	1.00	1.00	1.00	1.00

93

Example 3 This last example uses again the standard knots, as given above but now with the smallest number of these possible, that is covering one interval. This requires the use of eight knots (n = 3) and produces four basis functions. The knots are:

$$t_0 = t_1 = t_2 = t_3 = 0$$
$$t_4 = t_5 = t_6 = t_7 = 1$$

The four basis functions are now found using the algorithm for a general value of t $(0 < t < 1)$:

$p_0(t)$

$$
\begin{array}{cccc}
0 & & & \\
& 0 & & \\
0 & & 0 & \\
& 0 & & (1-t)^3 \\
0 & & (1-t)^2 & \\
& (1-t) & & \\
1 & & &
\end{array}
$$

$p_1(t)$

$$
\begin{array}{cccc}
0 & & & \\
& 0 & & \\
0 & & (1-t)^2 & \\
& (1-t) & & 3t(1-t)^2 \\
1 & & 2t(1-t) & \\
& t & & \\
0 & & &
\end{array}
$$

$p_2(t)$

$$
\begin{array}{cccc}
0 & & & \\
& (1-t) & & \\
1 & & 2t(1-t) & \\
& t & & 3t^2(1-t) \\
0 & & t^2 & \\
& 0 & & \\
0 & & &
\end{array}
$$

$p_3(t)$

$$
\begin{array}{cccc}
1 & & & \\
& t & & \\
0 & & t^2 & \\
& 0 & & t^3 \\
0 & & 0 & \\
& 0 & & \\
0 & & &
\end{array}
$$

It is seen that the B-spline basis functions produced here are precisely the Bézier basis functions. In this way we can regard cubic B-splines as being an extension of Bézier cubic curves.

B-splines and some of their properties

With the results of the last section we have now armed ourselves with a number of B-spline basis functions defined by a number of knots. For any integer $n > 3$, we have $n + 1$ basis functions and we choose $n + 1$ control points. We use the $n + 5$ knots introduced in the last section which are integer values over the interval 0 to $n - 2$. The notation is:

Basis functions $p_0(t), p_1(t), \ldots, p_n(t)$

Control points $r_0, r_1, \ldots, r_n$

Knots $\qquad t_0 = 0, t_1 = 0, 0, 0, 1, 2, 3, \ldots$

$$\ldots, n-4, n-3, n-2, n-2, t_{n+3} = n-2, t_{n+4} = n-2$$

The curve segment form we consider is that given by the following combination of the control points:

$$\mathbf{r}(t) = \sum_{j=0}^{n} p_j(t) \, \mathbf{r}_j \qquad t_0 \leqslant t \leqslant t_{n+4} \qquad (3.7)$$

As usual as t varies between $t_0 = 0$ and $t_{n+4} = n - 2$ the position vector $\mathbf{r}(t)$ changes and a curve is traced out. We refer to that part of the curve formed as t goes between t_i and t_{i+1} as a span of the curve. Thus this above example has $n + 4$ spans although some of these are degenerate due to the repeated knots at the beginning and ends of the range of knot values. Figure 3.19 shows an example of such a B-spline curve together with its control points.

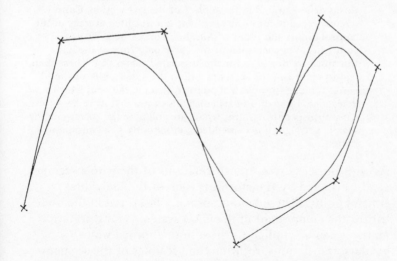

Figure 3.19 *A cubic B-spline segment and its control polygon*

95

B-spline curves have a number of properties. These are similar to those of Bézier curves and indeed as we saw in the last section we can regard B-splines as an extension of Bézier cubics.

- Since p_0 is the only one of the basis functions which is non-zero at $t = 0$, in fact $p_0(0) = 1$, the segment starts at the first control point r_0.
- In the same way the segment ends at $t = t_{n+4} = n - 2$ at the last control point.
- The functions p_0 and p_1 are the only ones to have non-zero derivatives at $t = 0$ and as a consequence of this and the values they take, the initial tangent direction is along the line from r_0 to r_1.
- Similarly, the tangent direction as the segment ends is along the line joining the last two control points.
- For t between t_i and t_{i+1}, precisely four of the basis functions are non-zero; these are p_i to p_{i+3}. It follows that the ith span of the curve segment is influenced only by the control points r_i to r_{i+3}.
- The converse of this is that any control point r_i affects exactly four of the spans of the curve; in fact these are the four consecutive spans formed as t goes between t_i and t_{i+4}. This provides us with the idea of local control; by moving any one of the control points we change only a part of the shape of the segment. An example of this is shown in Figure 3.20 where a single control point is moved to various positions and the corresponding effects upon the segments are plotted.
- The sum of the basis functions is unity (cf. Faux and Pratt, 1979); this is demonstrated in Example 2 of the last section. Consequently, each point on the segment is a weighted average of the control points and it lies within their convex hull. We can go further. For each span of the curve only four of the basis functions are non-zero for the associated values of t. These then add to unity and the span lies within the convex hull of the corresponding four control points which are the ones which determine the span. In particular this shows that if we have four consecutive control points which are collinear the corresponding span is a straight line embedded continuously in a continuous curve.

As with Bézier curves, transformations of the whole segment can be effected by transforming each of the individual control points. These would normally be stored in the node list for the component in the CAD system. Transformation matrices can be applied to them in the normal way to produce translations, rotations and scalings of the segment they define.

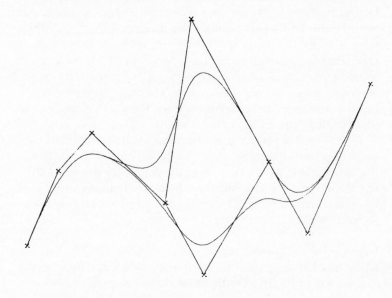

Figure 3.20 *Effect of changing a control point of a cubic B-spline segment*

Extensions of B-spline ideas

Although we have spent some time here discussing B-splines, we have only really looked at one aspect of them. It is possible to extend the ideas in a number of different ways.

The examples given above have all been based upon a particular form of selection for the knots. These have been equally spaced and the B-splines generated in this way are called uniform. By choosing them in different ways non-uniform B-splines are produced. This gives extra degrees of freedom in the design of the final curve shape. For the user of a CAD system who does not want to become involved in the details of the way in which his curves are defined, there should always be a means for selecting the knots and their spacing automatically. One way of choosing them in an automatic non-uniform way is to relate the intervals between the knots to the distances between the control points (Gordon and Riesenfeld, 1974).

Closed curves can be produced using B-splines. This is done by letting the control points start to repeat themselves;

we can also avoid having to repeat the knot values at the beginning and end. If we have n control vectors r_0 to r_{n-1} and we take the knots to be simply the integers 0, 1, 2, ..., then we can apply Equation (3.7), together with the Cox-de Boor algorithm for generating the basis functions, to form a closed curve; we need to interpret r_{n+1} to be the same as r_i whenever necessary. In this case, as the repetition of the knots is removed, there is no need for the curve generated to pass through any of the control points.

If the control points r_i in Equation (3.7) are replaced by vectors in homogeneous coordinates, we obtain the rational B-spline curves. Again there is extra flexibility provided by the ability to select and modify the homogeneous coordinate of each control vector. The properties of ordinary B-splines go over to the rational form.

Joining curve segments together

While smallish segments are fairly easy to define, they very often need to be fitted with other shapes to produce complete drawings or models of engineering components. We look at ways of joining together segments so that a sufficient degree of continuity is preserved across the joins. The ideas discussed in this section apply mainly to Bézier segments. They do work for B-spline curves, but it is possible to piece together B-spline segments by changing the knot values of one so that they follow on from the ones for the other, and then merely merging the two sets of knots, and of control points. This produces a single curve segment which reproduces the combined shapes of the original two.

Although we talk about joining segments together, we cannot really work in this way. It is impossible to design segments separately and then hope that in some automatic way they will fit properly together. In fact we are really looking at ways for designing large smooth curves that are composed of a number of Bézier (or other) segments. We consider the case of ordinary Bézier cubics and use the notation that for the ith segment the control points have position vectors a_i, b_i, c_i and d_i; the first and last here are the end-points of the segment and the other two are the tangent control points. With care the ideas discussed go across to the higher order Bézier curves and to rational ones.

For basic continuity, the end of one segment should be the

same as the start of the next. Here the advantage of the Bézier formulation is apparent since we can immediately write down a condition for this in terms of the control points. It is:

$$\mathbf{d}_i = \mathbf{a}_{i+1} \qquad \text{for all relevant i}$$

For smoothness across the join we need the tangent to be continuous there. Continuity of the tangent direction is achieved by ensuring that the end-tangent directions are the same; that is $(\mathbf{d}_i - \mathbf{c}_i)$ and $(\mathbf{b}_{i+1} - \mathbf{a}_{i+1})$ must be parallel, and this is equivalent to the three points $\mathbf{c}_i$, $\mathbf{d}_i = \mathbf{a}_{i+1}$ and $\mathbf{b}_{i+1}$ being collinear. Thus at a simple interactive level a user can ensure that segments join together smoothly by manipulating the control points so that this collinearity condition is fulfilled.

In the next section we apply these ideas to forming curves composed of a large number of segments which pass through a given number of points.

Curves through given points

Perhaps the most natural way for a CAD system to give a user access to smooth curves is to allow him to specify a number of points and then automatically to put a smooth curve through them. In this way the user does not need to know about Bézier curves, control points and the like and can simply press on and use the shapes that the system generates, altering them by redefining some of the specified points (or defining more) if the shape produced is not satisfactory.

In this section we look at two methods for putting a sequence of ordinary Bézier cubic segments together to define a large smooth curve through a number of given points. These points are denoted by $\mathbf{r}_1, \ldots, \mathbf{r}_n$. The first of these aims to provide continuity of the curve and its tangent only; the second ensures that the curvature is continuous. In both cases a single Bézier segment is placed between each consecutive pair of given points.

Method 1 Each segment goes between a pair of known points. So we know two out the four control points. It only remains to select suitable points for the tangent control points. This in turn depends upon the selection of a suitable tangent direction at each of the given points. We indicate one

99

way to do this which applies except for the points at either end of the list of given points. Take the point r_i and the ones r_{i-1} and r_{i+1} on either side of it. Through these three we interpolate a temporary quadratic curve. This could be an ordinary Bézier quadratic; we know control points for the ends and we can select the third one by making sure that the segment passes through the other given point r_i when the parameter is 1/2. Having fitted this temporary curve we can now obtain a suitable direction for the tangent at the given point r_i (by evaluating the derivative of the Bézier quadratic at 1/2), and we now discard the curve. Thus we establish a tangent direction at each of the given points except r_0 and r_n. The tangents here have to be selected by other means. Perhaps the curve is to join up smoothly with other existing parts of the CAD model, or the user may be asked directly to select a direction. Now for each pair of points r_i and r_{i+1} we can fit a Bézier cubic segment provided we pick the tangent control points somewhere along the end-tangent directions. This can be done for example by taking each to be at a distance d/3 from its appropriate end where d is the distance between the ends of the segment. Now the tangent points for consecutive segments are collinear with the common end-point of the segments and the condition for continuity of tangent direction is achieved.

Method 2 Here we ensure that there is continuity of curvature as well as of tangent direction. This is necessary for applications where smoothness is of critical importance, for example in the design of the surfaces of a car body. The eye can easily spot, partly from the way in which light is reflected, any sudden changes in curvature and as a result the design may not be aesthetically pleasing. (Some would even claim that if the curvature is continuous, it is also possible to detect sudden changes in the third derivative terms of a curve or a surface.) The curvature of curve segment $r(t)$ is in fact given by:

$$\kappa = |\dot{r} \wedge \ddot{r}| / |\dot{r}|^3$$

although we do not actually use this explicitly. Instead we go for a certain amount of overkill and arrange for continuity of r and its first two derivatives at each join of the curve segments. Extending the conditions obtained in the last section, this requires that the following hold:

$$\mathbf{d}_i = \mathbf{a}_{i+1}$$

$$\mathbf{d}_i - \mathbf{c}_i = \mathbf{b}_{i+1} - \mathbf{a}_{i+1}$$

$$\mathbf{d}_i - 2\mathbf{c}_i + \mathbf{b}_i = \mathbf{c}_{i+1} - 2\mathbf{b}_{i+1} + \mathbf{a}_{i+1}$$

If these relations are rearranged for segments $i-1$, i and $i+1$, then we obtain the following result which relates the unknown tangent points $\mathbf{b}_i$ to the known segment end-points $\mathbf{a}_i = \mathbf{r}_i$:

$$\mathbf{b}_{i-1} + 4\mathbf{b}_i + \mathbf{b}_{i+1} = 4\mathbf{a}_i + 2\mathbf{a}_{i+1}$$

Values for the vectors $\mathbf{c}_i$ can be deduced once the $\mathbf{b}_i$ are known since the continuity relations imply that:

$$\mathbf{c}_i = 2\mathbf{a}_{i+1} - \mathbf{b}_{i+1}$$

Now the above relation in the $\mathbf{b}_i$ represents a set of simultaneous equations for finding these tangent control points. Again we need extra information about the tangent directions at $\mathbf{r}_0$ and $\mathbf{r}_n$, the extreme ends of the curve. If the curve is to be chosen, this information is not needed and we arrive at a matrix equation:

$$\begin{bmatrix} 4 & 1 & 0 & 0 & \cdots & 0 & 1 \\ 1 & 4 & 1 & 0 & \cdots & 0 & 0 \\ 0 & 1 & 4 & 1 & \cdots & 0 & 0 \\ \vdots & \vdots & \vdots & \vdots & & \vdots & \vdots \\ \vdots & \vdots & \vdots & \vdots & & \vdots & \vdots \\ 1 & 0 & 0 & 0 & \cdots & 4 & 1 \end{bmatrix} \begin{bmatrix} b_{1x} \\ b_{2x} \\ b_{3x} \\ \vdots \\ \vdots \\ b_{nx} \end{bmatrix} = \begin{bmatrix} 4a_{1x} + 2a_{2x} \\ 4a_{2x} + 2a_{3x} \\ 4a_{3x} + 2a_{4x} \\ \vdots \\ \vdots \\ 4a_{nx} + 2a_{1x} \end{bmatrix}$$

for the x-coordinates of the $\mathbf{b}_i$; similar equations hold for the y- and z-coordinates.

Perhaps a warning should be given at this point. We have here arranged for continuity of $\mathbf{r}(t)$ and its derivatives across each segment boundary. However these are derivatives with respect to the parameter t which we have been assuming runs between 0 and 1. However the parameter does not influence the curve shape (it only indicates a speed at which it is traced out) and a change in the parameter would result in different values for the derivatives used. The technique described above produces satisfactory results provided the distances between pairs of given points $\mathbf{r}_i$ and $\mathbf{r}_{i+1}$ are all approximately the same. An alternative parametrization would be in terms of length along the segment. However as it is these segments we are trying to determine we do not know this before we start; we might instead use the 'chord length', that is the distance between the pairs of given points, and arrange for the parameter for the segment to range from zero to this value (instead of one).

CHAPTER 4

Intersection of curves

Introduction

When straight lines and curves are created on a CAD system it is extremely unlikely that they are immediately of the correct form. The user has to ensure that the various parts of the geometry fit together correctly. Sometimes he will be including entities merely for constructional purposes as one would on a conventional drawing board. For example, it may be required to blend two circular arcs together to form a new curve. The final form cannot be specified immediately. It is necessary to construct both arcs in full, find where their intersection is and then trim away the parts that are not required. On a drawing board the intersection is found by eye (and the trimming performed with an eraser). To be able to do the same thing on a CAD system requires the presence of commands that find the intersection of specified geometric entities. The trimming is easily accomplished by redefining an end of each entity to be the calculated point of intersection and then removing the unwanted parts from the display (possibly by redrawing it).

In this chapter we look at the intersection of geometric entities. Often this involves the solution of non-linear equations by iterative means. Various tests can be used to speed up these types of algorithms and some of these are discussed. We begin with the easiest of entities to deal with — straight line segments.

Intersection of straight line segments

If we have two straight lines in a plane, then they can be described by equations of the form:

$$a_i x + b_i y + c_i = 0$$

It is better to use this form rather than one such as $y = mx + c$

since it permits the representation of lines with infinite gradient by taking the b coefficient to be zero. There is a certain ambiguity about the choice of coefficients since multiplication of each of the three by any (non-zero) scalar yields a different equation describing the same line. If we need to we can remove this ambiguity by insisting that the sum of the squares of the coefficients should be unity (and maybe that the first non-zero coefficient should be positive).

The two equations for the straight lines can be regarded as a pair of simultaneous equations for finding x and y. It is easy to write down formulae for the solution to the equations merely by eliminating one of the unknowns and finding the other. The solution values are the x- and y-coordinates of the point of intersection of the two lines. The only situation where things can go wrong is if the lines are parallel (or coincident). We can test for this by checking the value of the determinant of the coefficients, that is the expression $a_1 b_2 - a_2 b_1$; if it is zero, then the special case exists and no single intersection point can be found.

In practice we need to be just a little more careful. We normally are dealing with line segments. If we use the node list scheme for storage, then we are holding the end-points. The segment that goes between the points (x_1, y_1) and (x_2, y_2) is part of the infinite line whose equation (in the above form) is:

$$(y_2 - y_1)x - (x_2 - x_1)x + (y_1 x_2 - y_2 x_1) = 0$$

If another segment goes between (x_3, y_3) and (x_4, y_4), then the two are parallel if:

$$(y_2 - y_1)(x_4 - x_3) - (x_2 - x_1)(y_4 - y_3) = 0$$

However, because we cannot store real number values exactly within a computer system and errors creep in due to the inexactness of the arithmetic, we should test that the above expression has a small absolute value rather being exactly zero. The meaning of 'small' depends upon the scale of the values being input. If the segments are found to be parallel we can see if they are coincident by substituting the coordinates of an end-point of one into the equation of the other; if the result is small, then there is coincidence.

When the segments are not parallel, we can proceed to find the intersection as described previously. This is the intersection of the two infinite lines that contain the segments.

Very often we are only interested in the case where the intersection point actually lies on both of the segments. (There are situations when intersections not on the segments are useful for defining new points for the construction of entities and some CAD systems allow such points to be handled.) So we need to check on the position of the inter-section point found. If it has coordinates (x, y), then it lies on the first segment if x is between x_1 and x_2, or equivalently if y is between y_1 and y_2. Because of the possibility of arithmetic errors, we may need to allow the point even if its coordinates lie just outside the acceptance range.

The idea of testing against a range of coordinate values can be used to avoid unnecessary computation. For example, if x_1 and x_2 are both less than x_3 and x_4, then the segments cannot possibly overlap (cf. Figure 4.1). In this case we need

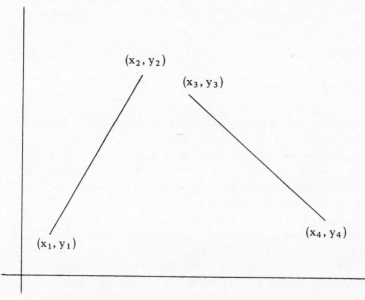

Figure 4.1 *Two non-intersecting line segments*

not continue with the computation. In fact we can avoid it if any of the following hold:

$$\max(x_1, x_2) < \min(x_3, x_4)$$
$$\min(x_1, x_2) > \max(x_3, x_4)$$
$$\max(y_1, y_2) < \min(y_3, y_4)$$
$$\min(y_1, y_2) > \max(y_3, y_4)$$

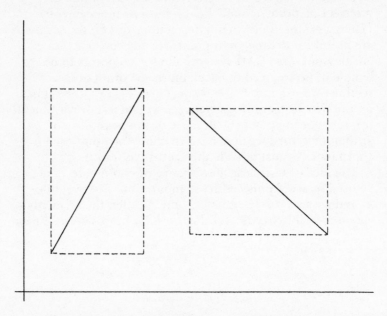

Figure 4.2 *Non-overlapping boxes around line segments*

Essentially what we are doing here is to put a box around each segment; if the boxes do not overlap, then the segments cannot intersect. This is illustrated in Figure 4.2. We call this type of procedure 'box-testing'. There are cases where the boxes overlap but the segments still miss each other and one is shown in Figure 4.3. For this reason it may be debatable whether box-testing really saves anything in this case. However as we see later on it is a technique that can be applied effectively for curves.

When the segments we consider are in three-dimensional space we can run into problems. We can apply the box-testing procedure if we want; on this occasion the boxes are cuboids in space with sides parallel to the coordinate planes and we need to introduce a test on the z-coordinate value of the end-points of the segments. The main problem is that the lines may not intersect and may not be parallel; they may be skew. Mathematically our difficulty is that each line provides us with two equations, so that we have four equations in all but only three unknowns. Thus we have an overdetermined problem.

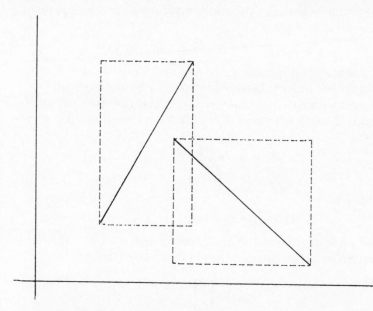

Figure 4.3 *Overlapping boxes around non-intersecting segments*

It is perhaps better to introduce parameters. Thus with obvious notation, typical points on the two segments have coordinates:

$$x = x_1 + (x_2 - x_1)u \qquad y = y_1 + (y_2 - y_1)u \qquad z = z_1 + (z_2 - z_1)u$$
$$x = x_3 + (x_4 - x_3)v \qquad y = y_3 + (y_4 - y_3)v \qquad z = z_3 + (z_4 - z_3)v$$

or in terms of their positions vectors we have:

$$r = r_1 + (r_2 - r_1)u$$
$$r = r_3 + (r_4 - r_3)v$$

where u and v are parameters between 0 and 1. Equating corresponding coordinate expressions gives three equations for finding the two unknown parameter values. One approach is to solve two of these equations and to check the solution in the third. If it fits, then the lines intersect (though not perhaps the segments) and the intersection point can be found by substituting one of the parameters into the relevant coordinate expressions.

An alternative technique which has the advantage of preserving symmetry among the coordinates is to consider

107

the distance between the above typical points. The square of this distance is given by:

$$S = |\mathbf{r}_3 - \mathbf{r}_1 + (\mathbf{r}_4 - \mathbf{r}_3)v - (\mathbf{r}_2 - \mathbf{r}_1)u|^2$$

and this has a minimum value as u and v vary of zero if the two lines intersect. Forming the partial derivatives with respect to the two parameters and equating these to zero yields the following pair of simultaneous equations for u and v:

$$\mathbf{d}_1 . \mathbf{d}_1 u - \mathbf{d}_1 . \mathbf{d}_2 v = \mathbf{d}_1 . (\mathbf{r}_3 - \mathbf{r}_1)$$
$$-\mathbf{d}_1 . \mathbf{d}_2 u + \mathbf{d}_2 . \mathbf{d}_2 v = \mathbf{d}_2 . (\mathbf{r}_3 - \mathbf{r}_1)$$

where:

$$\mathbf{d}_1 = \mathbf{r}_2 - \mathbf{r}_1 \quad \text{and} \quad \mathbf{d}_2 = \mathbf{r}_4 - \mathbf{r}_3$$

The form of this matrix equation is a little clearer if it is written in matrix form. It can be rearranged as:

$$A^T A \begin{bmatrix} u \\ v \end{bmatrix} = A^T \mathbf{b}$$

where:

$$A = \begin{bmatrix} x_2 - x_1 & -x_4 + x_3 \\ y_2 - y_1 & -y_4 + y_3 \\ z_2 - z_1 & -z_4 + z_3 \end{bmatrix} \qquad \mathbf{b} = \begin{bmatrix} x_3 - x_1 \\ y_3 - y_1 \\ z_3 - z_1 \end{bmatrix}$$

In matrix form the original three equations in the two unknowns look like:

$$A \begin{bmatrix} u \\ v \end{bmatrix} = \mathbf{b}$$

Thus we have a 'least squares' rearrangement of the original overdetermined system as is discussed in Chapter 1.

This type of optimization strategy is typical of a number of methods for dealing with intersections of curves and surfaces. They often involve finding the minimum of a positive quantity which is usually the square of some relevant distance. Their advantage is that information can always be gained. In the case of the two line segments, even if we find that they do not intersect in three-dimensional space, we have still succeeded in finding their minimum distance apart.

Non-linearity and the intersection of curves

In this section we investigate the possibility of applying the
approach used above for finding the intersection of two line
segments to the problem of finding the intersection of two
curves. This presents certain problems since the equations we
have to solve are now non-linear and require the use of
iterative techniques. We begin with a simple case — that of a
line and a circle.

Straight lines and circular arcs are the most common
entities to appear in engineering drawings. Some simple
two-dimensional draughting systems in fact have these as
essentially their only geometric entities. Intersections
between them are reasonably straightforward to deal with.
The equation of a line is given by:

$$ax + by + c = 0$$

and that of a circle with centre at (p, q) and radius r is:

$$x^2 + y^2 - 2px - 2qy + (p^2 + q^2 - r^2) = 0$$

If we assume that a is non-zero, we can find x from the line
equation, substitute it into the circle equation and obtain a
quadratic equation for y. This can be solved by the well-
known formula. If it has real roots, then the line cuts the
circle in two points; if it has non-real roots, then the line
misses the circle; and if the roots are equal, then we have a
case of tangency. If we wrote a high-level language program
to perform this sort of calculation there would apparently be
no need for extensive iterative calculation. But it is there
nonetheless. We require the evaluation of a square root
to solve the quadratic and the compiler would insert an
appropriate iterative routine to do this for us.

The line and circle illustrate some of the problems that
are present with general curve intersections. The two curves
may or may not intersect; there may be more than one
intersection point; or there may be instances of coincident
intersections as in the case of tangency. It happens that the
geometry of the line and circle are well understood and they
are sufficiently specific curves that we know how to allow for
these difficulties. The problems magnify when we deal with
more general curve forms.

In cases when there is more than one intersection between
entities, it is usual in CAD systems for the user to indicate

which one he wishes to use. For example, when specifying the intersection of a circle and a line, say, the user will usually select these entities by pointing to them on the graphics display by a light pen or via an electronic tablet. The simplest procedure is then to take the point of intersection nearest to one of the points used to pick the entities.

While dealing with the circle, we note in passing that the intersection of two circular arcs can be reduced to the above case by considering the common chord. If the equations of the circles have the form given above with suitably chosen subscripts, then the equation of the common chord is formed by taking the difference:

$$2(p_2 - p_1)x + 2(q_2 - q_1)y - k = 0$$

where:

$$k = p_2{}^2 - p_1{}^2 + q_2{}^2 - q_1{}^2 - r_2{}^2 + r_1{}^2$$

The circles intersect at the same points as this line.

We now turn to more general curves. Attention is restricted initially to plane curves. We consider two planar curves in parametric form, $r_1(u)$ and $r_2(v)$ where u and v are parameters (going between 0 and 1). In order to find their intersection we need to solve the vector equation:

$$r_1(u) - r_2(v) = 0 \qquad (4.1)$$

This represents two non-linear equations in two unknowns. We apply the Newton-Raphson method which is discussed in Chapter 1. If $d(u, v)$ denotes the above difference of the two position vectors, then for small changes Δu and Δv in the two parameters, the change made in d is given by the first terms of a Taylor expansion:

$$d(u + \Delta u, v + \Delta v) = d(u, v) \quad \frac{\partial d}{\partial u} \Delta u + \frac{\partial d}{\partial v} \Delta v$$

If $u + \Delta u$ and $v + \Delta v$ represent a better solution to Equation (4.1) than u and v do, then we approximate the left hand side of the last equation to zero and obtain an equation for finding u and v. Rewriting d and its partial derivatives in terms of the original position vectors we obtain:

$$\begin{bmatrix} \dfrac{\partial r_1}{\partial u} & -\dfrac{\partial r_2}{\partial v} \end{bmatrix} \begin{bmatrix} \Delta u \\ \Delta v \end{bmatrix} = -r_1(u) + r_2(v) \qquad (4.2)$$

The first term in the square brackets here is a 2×2 matrix whose columns are the column vectors shown.

The way this scheme is applied is first to select starting values for the parameters u and v which should preferably be near the actual values at the intersection. Equation (4.2) is used to find Δu and Δv and these are added to the current values of u and v. The changes are then recalculated, u and v updated and the process continues until convergence is achieved. This can be tested for by looking at the absolute values of Δu and Δv found at each stage; when these are sufficiently small the iteration should be stopped.

Several problems can arise and we look at these in turn. First, we need good starting values. While in principle any reasonable values will do, in practice the closer they are to the solution values the fewer steps are needed and the more likely is convergence to be achieved. If no other information is at hand, a possible method is to look at the intersection of the chords between the end-points of each of the two segments. This is a straight line problem and the intersection can be found easily. The estimate for each of u and v is then the proportion of the way along the appropriate chord that the intersection lies.

If the curve segments do not intersect, then naturally the iterative scheme does not converge. This might manifest itself in different ways; perhaps the values of u and v will become large and tend towards infinity (or at any event go outside the acceptable range of 0 to 1), or maybe they will merely wander about in an apparently random way. This latter case is more difficult to detect and deal with. In any event an upper limit should be placed upon the number of iterations that are allowed and the segments declared not to meet if convergence has not been achieved by then.

The values of u and v obtained during the iteration process may go outside the permitted interval of 0 to 1. This could be a purely temporary phenomenon, possibly caused by a bad choice of start point, and they will come back into the range after a few more steps. This means that we must not be too quick about stopping the iterations if the results seem to be going astray.

Finally, the above scheme has some difficulty in picking up more than one intersection. If two or more exist, then the one that is found is usually the one nearest the start point;

but this is by no means guaranteed. It might be possible then to select different start points to find any others. However if we are looking for automatic ways to find all intersections this approach is of little use if we do not know how many we have to find. In interactive operations it is possible to allow the user to intervene, but this is hardly satisfactory.

In the next section we look at other techniques which can be used together with the iterative scheme to eliminate some of these difficulties.

We now give an example of the iterative scheme used to find the intersection of the following pair of Bézier cubics. They are shown in Figure 4.4.

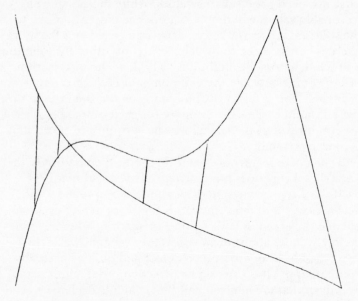

Figure 4.4 *Intersection using Newton-Raphson iteration*

$$\mathbf{r}_1(u) = \begin{bmatrix} 0 \\ 0 \end{bmatrix}(1-u)^3 + 3\begin{bmatrix} 2 \\ 10 \end{bmatrix}u(1-u)^2 + 3\begin{bmatrix} 4 \\ -2 \end{bmatrix}u^2(1-u) + \begin{bmatrix} 8 \\ 8 \end{bmatrix}u^3$$

$$\mathbf{r}_2(v) = \begin{bmatrix} 0 \\ 8 \end{bmatrix}(1-v)^3 + 3\begin{bmatrix} 1 \\ 2 \end{bmatrix}v(1-v)^2 + 3\begin{bmatrix} 6 \\ 2 \end{bmatrix}v^2(1-v) + \begin{bmatrix} 10 \\ 0 \end{bmatrix}v^3$$

We start the iterations for this example with both u and v set to unity. The following table shows the values obtained for the parameters over the first nine iterations.

Step	u	v
0	1.0000	1.0000
1	0.8056	0.6389
2	0.5995	0.5021
3	0.0957	0.1489
4	0.2118	0.2423
5	0.2645	0.2751
6	0.2745	0.2820
7	0.2748	0.2823
8	0.2748	0.2823
9	0.2748	0.2823

It is seen that the process has converged over these steps. The coordinates of the common point on both curves given by the appropriate limits of the parameter values are $x = 1.6906$ and $y = 4.1732$. To show the convergence process graphically, Figure 4.4 also shows lines joining points on the two curves given by the parameter values at each stage of the iteration. Although convergence does occur, it is clear from the figure that the starting values used are not very well selected. This was done in order to illustrate a problem that can occur. It is seen that at step 3 the parameters suddenly jump and represent points on the 'other side' of the intersection. Indeed the u value very nearly becomes negative so that it no longer represents a point on the segment. Such behaviour can occur in the early steps of the iteration and can complicate tests for the convergence or divergence of the process.

The type of iterative technique used here for ordinary parametric curves can also be adapted to rational forms. There is need for some rearrangement in order to present the equations to be solved. If W denotes the homogeneous coordinate, then the intersection point has parameter values which are the solution of the following pair of simultaneous equations:

$$X_1(u)W_2(v) - X_2(v)W_1(u) = 0$$
$$Y_1(u)W_2(v) - Y_2(v)W_1(u) = 0$$

If the curve segments are not planar but exist in three-dimensional space, then we have the problem that we had for line segments, namely that we have more equations than unknowns. Again a good method of approach is that of finding the shortest distance between the two curves. If this is found to be zero, then the curves meet and the point where the shortest distance occurs is the intersection

113

point. For non-rational forms this means minimizing the expression:

$$|\mathbf{r}_1(u) - \mathbf{r}_2(v)|^2$$

For Bézier cubic segments this is a complicated expression of the sixth degree in u and v. The two partial derivatives are zero at a minimum value and these provide two non-linear equations for u and v which we can attempt to solve by the Newton-Raphson scheme or one of its refinements. This is aided by the fact that the Bézier form gives immediate access to the derivatives of $\mathbf{r}_1$ and $\mathbf{r}_2$. For more complicated forms when the derivatives are not straightforward to obtain analytically, we can attempt the minimization problem by one of the methods of optimization theory (cf. Walsh, 1975). For example, a direct search method could be used, evaluating the function at several values of u and v and using the results obtained to decide which values to try next. This continues until it is clear that no smaller value of the function can be obtained.

Some of these difficulties are again reduced by the subdivision techniques we investigate in the next section.

Subdivision and box-testing techniques

In the last section we tried to apply the techniques for intersecting two line segments to two curve segments. However we have difficulties in deciding whether or not the segments really do intersect and if so in how many points. The problems stem from the fact that if the curves are long and fairly twisted they do not resemble line segments and the techniques used cannot be expected to work well.

If we rewrite Equation (4.2), then we see that it takes the form:

$$\mathbf{r}_1(u) + \frac{\partial \mathbf{r}_1}{\partial u} \Delta u = \mathbf{r}_2(v) + \frac{\partial \mathbf{r}_2}{\partial v} \Delta v$$

As Δu and Δv vary, the two sides of this expression trace out tangents to the two curves at the points corresponding to u and to v. Thus we are essentially trying to intersect the two tangents. We should not expect this approach to be successful unless the curve is a good approximation to its own tangent. We can improve matters, however, by subdividing each full segment into smaller subsegments. Each of these should be

small enough that it is reasonably close to being straight.
By trying to find the intersection of these in pairs, we can
find the intersections of the two original segments. Furthermore, we are more likely to find multiple intersections
when they exist; if we have split the curves so that different intersections occur in different pairs of subsegments,
then each should be found when the subsegments are
intersected.

However consider the amount of work required for this
procedure. If each original curve is divided into five parts,
then there are 25 pairs of subsegments to examine for intersections. Each requires its own iterative scheme. If the
original ones only meet in one point, then 24 of these will
produce negative results and effort has been wasted. We need
a simply applied method for seeing quickly if subsegments
have a chance of intersecting.

This quick check is provided by box-testing. We looked at
this for line segments earlier in this chapter. A rectangle
(or in three-dimensional case, a cuboid) is placed around the
segment. Its sides are parallel to the coordinate axes. If the
boxes of two segments do not overlap, then the segments
cannot meet. The test for overlapping is simply a check on a
small number of inequalities. The boxes for line segments can
be obtained by inspecting the coordinates of the end-points.
This does not work for curve segments. The Bézier (or
B-spline) formulation is important here. This provides us with
a number of control points and the segment lies within the
convex hull of these points. If we then take the smallest box
containing all the control points, then the segment certainly
lies within it. An example of this is shown in Figure 4.5 for
a Bézier cubic segment.

All that remains now is to be able to obtain the control
points for curves that are subsegments of larger Bézier
curves. Such subsegments are determined by coordinates
which are cubic polynomials in the parameter; by a linear
change of variables the parameter can be made to run over
the range 0 to 1 and so the subsegment can be expressed in
Bézier form. The de Casteljau algorithm can be used here.
In the form given in Chapter 3, this can be used to split a
segment into two parts at the point with the parameter
value t. Slightly more generally, if the control points of
a Bézier cubic are a, b, c and d, then those of the subsegment

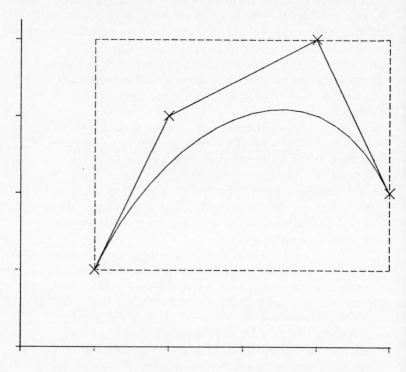

Figure 4.5 *Box constructed around control points for a segment*

going between points with parameter values s and t
$(0 < s < t < 1)$ are given by:

$$\mathbf{a}(1-s)^3 + 3\mathbf{b}s(1-s)^2 + 3\mathbf{c}s^2(1-s) + \mathbf{d}s^3$$
$$\mathbf{a}(1-s)^2(1-t) + 2\mathbf{b}s(1-s)(1-t) + \mathbf{b}(1-s)^2t + \mathbf{c}s^2(1-t) + 2\mathbf{c}s(1-s)t + \mathbf{d}s^2t$$
$$\mathbf{a}(1-s)(1-t)^2 + \mathbf{b}s(1-t)^2 + 2\mathbf{b}(1-s)t(1-t) + 2\mathbf{c}st(1-t) + \mathbf{c}(1-s)t^2 + \mathbf{d}st^2$$
$$\mathbf{a}(1-t)^3 + 3\mathbf{b}t(1-t)^2 + 3\mathbf{c}t^2(1-t) + \mathbf{d}t^3$$

Thus the box of the subsegment between s and t is the one
containing these points.

In fact we can find the intersection of two curves by using
the box-testing procedure alone without any need for
Newton-Raphson iteration. We merely keep subdividing
the two segments until the boxes obtained are smaller than
the acceptable error for the result. The flow diagram shown
in Figure 4.6 gives an algorithm for a procedure called
Box-Test with four arguments s_1, t_1, s_2 and t_2. This finds the

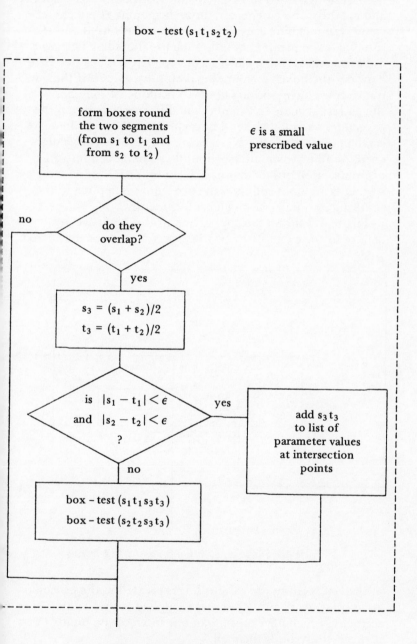

Figure 4.6 *Box-testing algorithm*

117

intersections of two subsegments of specified Bézier curves r_1 and r_2; the subsegments are those beginning at parameter value s_1 and ending at t_1. The procedure calls itself and so is suitable for implementation in such programming languages as Pascal and C which allow recursion. The basic action is to consider the boxes surrounding each subsegment. If they do not overlap, then nothing needs to be done. If they do, then a test is made to examine the differences between the parameter values at the end of each subsegment. If these are small enough, then the averages are taken to be the values corresponding to an intersection; these are added into a list of intersection points that have been found. Otherwise the segments are both split and the procedure is applied to the resulting four pairs of smaller subsegments.

Figure 4.7 shows the application of this recursive procedure to the example of two Bézier cubic curves considered

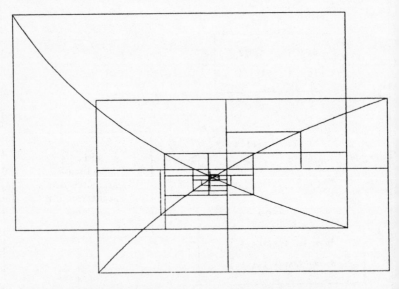

Figure 4.7 *Single intersection found by box-testing*

in the last section; the various boxes created in the process are displayed. Figure 4.8 shows it applied to two cubic segments which have more than one intersection point; it successfully finds them all.

Once the procedure has finished it is necessary to examine

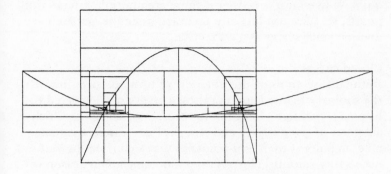

Figure 4.8 *Pair of intersections found by box-testing*

the list of intersections that have been found. It is possible
that the same one is recorded more than once. This happens
if an intersection happens to lie on the edge of a box at the
final stage as it is also likely to be on the edge of an adjacent
box at some stage of the process. If the two curves overlap
(or come very close) over some interval, then several points
along this will be found; it is then essential to set an upper
limit to the number of intersections that can be found and to
come out of the procedure when this is reached.

The two methods for intersections that we have looked at
are at opposite ends of a spectrum. The recursive subdivision
approach represents those methods whose basic step is
simple, just testing a few inequalities, but has to be repeated
a very large number of times. The Newton-Raphson scheme
is typical of methods where the basic step involves the
computation of complicated mathematical formulae, but
which has to be repeated less often (assuming convergence
is achievable). On the whole the subdivision approach is less
prone to error due to peculiarities in the curve form.

Compromise intersection methods can be devised. For
example, at each stage of a subdivision scheme we could use
the Newton-Raphson scheme to optimize the choice of the
next subdivision. If we apply one, or possibly two, iterations
we obtain a prediction of where an intersection is likely to
exist. We can then split each segment into three parts; the
middle one is chosen to be relatively small and to include the
predicted point of intersection. If the prediction is good,
then large amounts of the segments can be disposed of

quickly. Even if it is not, or if there are multiple intersection points, we have not lost any accuracy since the next subdivision stage serves as a correction.

Closed curves

In the last sections of this chapter we turn our attention to the subject of closed curves. These curves are considered to be planar and to be built up out of a set of segments which are parts of given curves and straight lines. These segments meet in pairs at their end-points so that a continuous (but not necessarily smooth) closed curve is obtained. Our reason for studying such curves is partly as a preparation for Chapter 6. But perhaps more importantly they serve as an important illustration of the fact that geometrical properties which are obvious to the human brain require sophisticated algorithms for their verification when this is performed computationally. These properties include such aspects as determining whether a given point lies inside or outside a closed planar curve and what constitutes the outer boundary of a region defined as the intersection of the areas of several overlapping curves.

Closed curves do have their own applications. Plotting times for drawings produced from CAD systems are improved if the geometric entities can be drawn consecutively without the pen leaving the plotting surface. Similarly NC machining efficiency is increased if the tool does not have to keep leaving the work-piece but can traverse the whole outer boundary in one passage. Often we need the area of a region defined on a CAD drawing or model of a component as being bounded by a number of geometric entities. These need to piece together correctly if the area is to be calculated by the system. The analogy in three dimensions of closed planar curves is that of surface faces of a geometric model of a component. It is important that these faces fit together precisely, for otherwise they do not truly define the volumetric region occupied by the component.

We consider first the problem of deciding whether a given point lies inside or outside a closed curve. Two possible methods are described. The first involves the construction of a semi-infinite line drawn from the point. Computationally, it is difficult to deal with the infinite and the line is actually drawn from the given point to some other point at least one of whose coordinates is much larger (or smaller) than those of points involved in the curve. We then need to find where this line intersects the curve and this involves testing it against each of the segments which form it. (During this process it can also be checked to see if the point actually lies on the curve itself.) The total number of intersections (which could be zero) is thus found. As Figure 4.9 shows, if this number is even, then the point lies outside the curve; if it is odd, then it is inside.

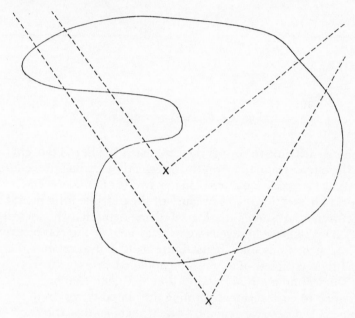

Figure 4.9 *Inside or outside a closed curve?*

This method is not foolproof. If the construction line passes through (or close to) a point where two segments meet, then we could find an intersection twice and so obtain the wrong total number. It is not even sufficient to throw away repetitions of intersection points; Figure 4.10 shows two examples, one where we ought to count a single inter-section with the construction line, the other where we need to count it twice. These ambiguities could be eliminated by examining the tangent directions of the segments at the intersection point. However this is time-consuming and it is simpler merely to choose another construction line and try again.

121

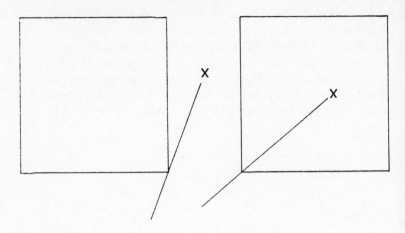

Figure 4.10 *Two awkward cases*

More difficult to spot is the case when the line is tangent to the curve. Depending on the intersection-finding procedure used, we may identify zero, one or two (or perhaps even more) intersections. It is because of the possible infrequent occurrences of such pathological phenomena that the user of a CAD system really needs some awareness of the complexity of the algorithms that are used internally and so retain a healthy scepticism of any result produced.

The second method requires that an orientation be assigned to each segment forming the closed curve. In the figures used here we generally orient them so that the curve is traced out anticlockwise. For each segment we evaluate the angle that its endpoints subtend at the given point we are testing. This evaluation can in fact be quite approximate. The orientation is needed so that the algorithm used to find the angles can be consistent and give negative results in certain cases. As Figure 4.11 shows, when the point lies inside the curve, the sum of the angles is $360°$; when it is outside the signs of the angles can be arranged so that the sum is zero. Thus we test the value of the sum and decide on the position of the point. Theoretically the sum should be $180°$ if the point lies on the curve itself; however it is likely that in such a case the angle evaluation algorithm will break down for the segment concerned or at least that it will provide the warning directly.

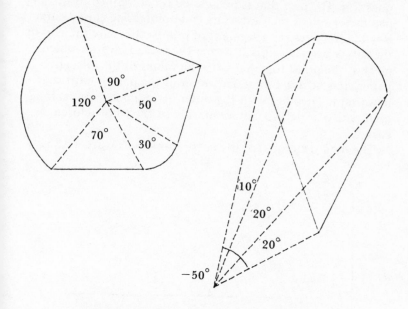

Figure 4.11 *Use of angles subtended at a point*

Thus there is a large amount of work involved within a CAD system in determining a relationship between two entities which is not immediately apparent from the graphics display to the user of the system.

A data structure for closed curves

The structure we describe here is an extension of the node and edge list idea discussed in Chapter 2. It is one that is suited to dealing with the intersection of closed curves which we discuss in the next section. Previously we have thought of an entity such as a line segment as being specified by two (or more) nodes. Here we additionally use the idea that a point can be specified by two entities; their intersection being the point in question.

It is important in CAD systems not to corrupt data more than is necessary. As we are going to be manipulating the various segments that form the closed curves we are studying, we aim to store these as sub-entities of entities·stored separately and not changed. We allow the possibility that a sub-entity may actually be the whole of the entity in question. The part that is being used is determined by storing parameter values relevant to its end-points. The closed curve is held effectively as a segment of points; these are the end of the sub-entities. Each is stored as the pair of entities which meet at the point together with the values of the relevant parameters at that point. The order of the pair of entities is important. Together with the order of parametrization along each entity, it defines an orientation of the whole closed curve.

The data structure for the curve shown in Figure 4.12 is

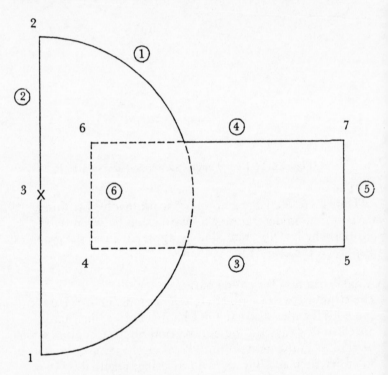

Figure 4.12 *An example closed curve*

given below. In the entity (edge) list, the third node number is used to specify the centre when the entity is a circular arc. Note that the two circular arcs shown are regarded as being parts of the same semicircular entity whose number is 1.

Node list				Entity list				Closed curve list				
No.	X	Y	No.	Type	Node 1	Node 2	Node 3	Ent 1	Ent 2	Par 1	Par 2	Next
1	0	−3	1	arc	1	2	3	2	1	1.00	0.00	2
2	0	3	2	lin	2	1		1	3	0.41	0.37	5
3	0	0	3	lin	4	5		4	1	0.63	0.59	4
4	1	−1	4	lin	7	6		1	2	1.00	0.00	1
5	6	−1	5	lin	5	7		3	5	1.00	0.00	6
6	1	1	6	lin	6	4		5	4	1.00	0.00	3
7	6	1										

We illustrate how to use the closed curve list by showing how it can be plotted out. Start with any entry, say the first. The point represented is at the intersection of entities with numbers 2 and 1. The parameter values for the point on each curve are held so we can find what the point is (if we need to). The important fact is that this entry shows that we are moving onto entity number 1 and the parameter value is zero. We now look through the closed curve list for entries whose first entity number is also 1 and select the one with the smallest value of the first parameter exceeding zero. This is the second entry in the list where entities number 1 and 3 meet. The value of the first parameter here is 0.41 and so we plot out that portion of entity number 1 defined between parameter values zero and 0.41. The value of the parameter associated with entity 3 is 0.37. We now look for entries whose first entity number is 3 and of these choose the one with the smallest entry for the first parameter which exceeds 0.37. This leads to the fifth entry and we plot out the part of entity 3 obtained as its parameter ranges between 0.37 and 1.00. Continuing in this way we produced the entire curve.

To avoid having to trace one's way through the structure every time the closed curve is to be used, it can be formed into a linked list. This is indicated by the last column, labelled 'next' in the structure above. This records the number of the entry within the list of the next point on the curve. Thus the 'next' entry for the first point is 2, and for the second point is 5. The above structure shows how these entries are completed.

The merging of closed curves
Suppose we are given two (or more) overlapping closed contours. By merging is meant the determination of the outside boundary of the region occupied by them both. The motivation here is the application discussed in Chapter 6

where the curves represent horizontal NC tool paths to produce certain predetermined shapes. In order to be able to produce a combination of such shapes we need to be able to remove those parts of the tool path where the shapes overlap. In this way the outer periphery of a complicated component can be generated.

Two overlapping curves are shown in Figure 4.13. An orientation is shown on the edges and as before we have taken this to be anticlockwise around each curve. If we know

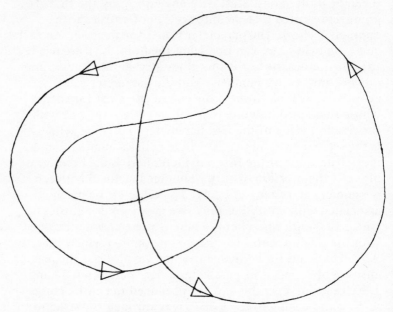

Figure 4.13 *Orientated closed curves*

a point on the outer boundary of the shape, we can pick up the rest of it by the simple procedure of tracking round the edges according to the direction of orientation and turning to the right whenever we come to an intersection. The concept of 'turning right' is simplified if we ensure that the edge segments are parametrized with the parameter value increasing in the direction of orientation. This approach is facilitated by the use of the data structure outlined above.

We are given two closed curves and these are stored in a closed curve list as above. It is necessary of course to find the intersections of the two curves. This involves checking each sub-entity of one with all those of the other. The box-testing approach will eliminate many of the cases. When an inter-section point is found we need to record it. It is determined by the entities on which it lies and the values of the relevant parameters on each. These can be entered into the closed curve list. The crux of the method is that we enter each intersection twice; the second entry has the order of the entities, and consequently the order of the parameters reversed. Figure 4.14 shows an example of this using the curve from the example of the last section and another; the data structure with the additions for the intersections is also shown.

A point on the outer boundary is now selected. This should be on the ends of the sub-entities stored in the original closed curve list. There are several ways of making the choice and to some extent it depends upon the precise way in which the various entities are defined. For example, it may be sufficient to take the segment end-point with the greatest x-coordinate value; or it may be necessary to consider points on one curve, test to see if they lie inside the other, and continue until one is found which lies outside.

Once a point on the outer boundary has been found, we simply apply, to the extended closed curve list, the pro-cedure for tracing out a curve described at the end of the last section. The construction of the list ensures that the outer boundary is obtained. In fact if we start at any point in the list, a closed curve is obtained. For the last example, three such curves are possible. These are shown in Figure 4.15. One is the outer curve. This represents the boundary of the union of the two original areas; the other two form the boundary of the intersection of these areas. If we are inter-ested only in the outer boundary, then the closed curve list should be purged of the unwanted intersection points.

This method has the advantage of avoiding some of the awkward cases of merging these types of curve. Such a case is when one curve just overlaps or just misses another. An example is shown in Figure 4.16 where a corner point of one curve is close to an edge of the other. Due to rounding errors in the arithmetic, we cannot be completely sure if the two edges at the corner node will be found to meet the other edge. Either both or neither could be found to cut it. The

127

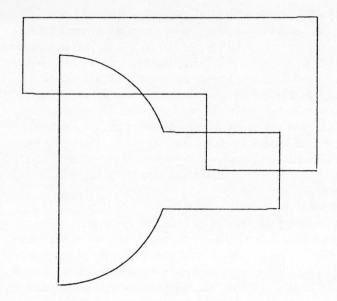

Closed Curve List

	ENT1	ENT2	PAR1	PAR2	NEXT	FLAG
1	2	1	1.000	0.000	2	1
2	1	3	0.414	0.366	17	1
3	4	1	0.634	0.586	13	1
4	1	2	1.000	0.000	15	−1
5	3	5	1.000	0.000	6	−1
6	5	4	1.000	0.000	19	−1
7	7	8	1.000	0.000	16	1
8	8	9	1.000	0.000	20	1
9	9	10	1.000	0.000	10	1
10	10	11	1.000	0.000	11	1
11	11	12	1.000	0.000	12	1
12	12	7	1.000	0.000	7	1
13	1	8	0.691	0.647	8	1
14	8	1	0.647	0.691	4	−1
15	2	8	0.167	0.200	14	−1
16	8	2	0.200	0.167	1	1
17	3	9	0.600	0.750	9	1
18	9	8	0.750	0.600	5	−1
19	4	9	0.400	0.250	18	−1
20	9	4	0.250	0.400	3	1

Figure 4.14 *Overlapping closed curves*

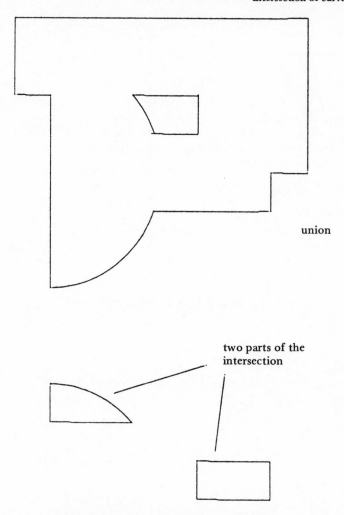

union

two parts of the
intersection

Figure 4.15 *Parts of the intersected curves*

method always produces an acceptable merged boundary.
Any of four outcomes is in fact possible and these are shown
in Figure 4.17 with the region of overlap drawn out in
exaggerated form. The small loops in two of the solutions
would normally be within the acceptable level of tolerance.
They could be removed by ensuring that the final curve is
always checked to eliminate edges of very small length (and
reconnect the curve).

129

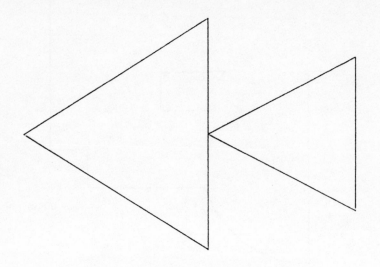

Figure 4.16 *Closed curves which touch*

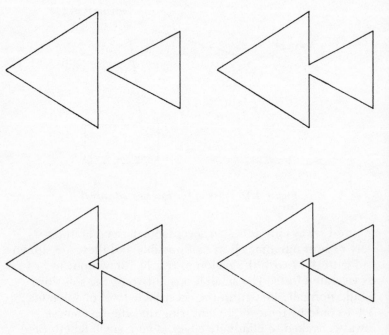

Figure 4.17 *Four possible intersection outcomes*

Closed curves can also be used to represent holes within
solid shapes. In this case the orientation of the edges should
be taken to be clockwise. Thus the area that is bounded
by the curve is always to one side (the left) as it is traced
out. Such holes complicate the derivation of the outer
boundary. One way to take account of them is to introduce
cuts across the bounded area to join together the inside
and outside boundaries. An example of this is shown in
Figure 4.18. Note that in amalgamating the boundaries in
this way, the orientations of the edges conform perfectly.

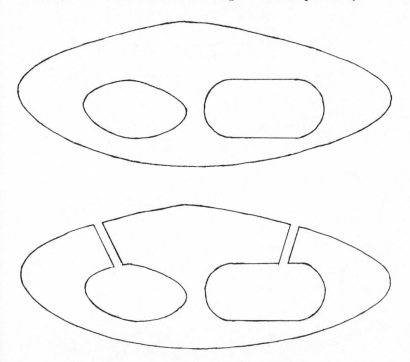

Figure 4.18 *Amalgamation to form a single boundary*

It is necessary to record in some way the fact that the extra
edges introduced to produce the cut are special; they need
to be removed when all merging is complete. With them
in place the boundary of each area is a single piece and the
previous algorithm can be applied. Figure 4.19 shows the
merging of two curves which bound areas containing several
holes.

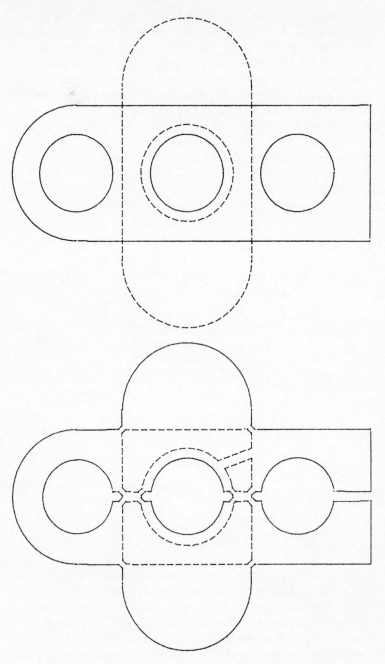

Figure 4.19 *Intersection of two complicated closed curves*

CHAPTER 5

Representation of surfaces

Introduction

In Chapter 3 we looked at some of the techniques for describing smooth curve shapes. Here we extend these ideas to look at smooth surfaces. The extension from an object, such as a curve, which is one dimensional (even though it is drawn in two- or three-dimensional space), to a surface which is two dimensional brings a number of extra difficulties which various developers in the field have attempted to overcome in various ways. The purpose of this chapter is mainly to examine some of the basic ideas and their inherent problems and so provide the reader with a familiarity with these without aiming to show what all the answers are. Indeed there are many areas of the use of surfaces in design that are not well understood and still present difficulties.

Commercial CAD systems do however exist which provide adequate facilities for surface definitions and their manipulations. These usually aim to provide the user with semi-automatic, interactive ways of dealing with surfaces. This approach serves to ease the burden for the user and to help to keep under control the confusions which would occur if free access to all the flexibility allowed by some definition techniques was allowed. As a consequence it is sometimes found that apparently simple operations on surfaces are denied the user. Nonetheless these systems are currently being used with success in industry as aids in the design and manufacture of such items as car bodies and moulds for complex plastic components.

We first discuss the familiar idea of ruled surfaces since these can be described mathematically fairly easily. Coons' patches are an extension of this idea. A patch is a part of a large surface usually with four bounding curves. It plays the same part for surfaces as do segments for curves. We then introduce Bézier and B-spline surface patches which provide flexibility in a form convenient for computational work and are often used in practice.

Ruled surfaces

A ruled surface is one that is generated by taking two curves and using straight lines to join together, in a consistent way, pairs of points, one from each curve. For example, we could take a circle (or any other plane curve) and form a copy by translating it in a direction perpendicular to its plane. If we then join corresponding points on both curves we obtain a cylinder. Another example for a different plane curve is shown in Figure 5.1. Here the corresponding pairs of points on the two curves are those with the same parameter values.

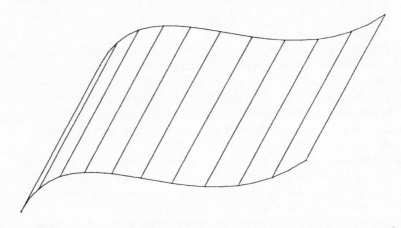

Figure 5.1 *A ruled surface*

A large number of everyday surfaces are ruled surfaces. They are well understood by engineers and as a consequence are used often and appear frequently in NC machining systems. They are also simple to describe from a mathematical point of view. Suppose that the two original curves are described by position vector functions of a single parameter u, namely $r_0(u)$ and $r_1(u)$. We assume that these curves are given in parametric form and that the parameter u goes between 0 and 1 to trace out the curve. Then for each value of u we can produce a point on each curve and we need to join these by a straight line. This is a matter of linear inter-polation and the resultant form for position vectors of points on the surface is:

$$r(u, v) = (1 - v)r_0(u) + vr_1(u) \qquad (0 < u, v < 1) \qquad (5.1)$$

In this equation, as u and v vary points in space are formed and we obtain 'a surface patch' such as that shown in Figure 5.1; the patch is the concept for surfaces which corresponds to the segment for curves. When v = 0 the right hand side of this expression reduces to $r_0(u)$ and when v = 1 it becomes $r_1(u)$. Thus the surface patch produced contains the two given curves on its boundary. For other constant values of v a curve which is a combination of these is described. When u is constant and v varies, a straight line in space is produced.

While this is the usual procedure for obtaining ruled surfaces in CAD systems, a word of caution needs to be given. We have relied upon the parametrization of the two initial curves to decide what is meant by 'corresponding points'. Normally this gives acceptable results. However consider the following two Bézier quartic curves:

$$r_1 = \begin{bmatrix} 0 \\ 0 \\ 0 \end{bmatrix}(1-u)^4 + 4\begin{bmatrix} 1 \\ 3 \\ 0 \end{bmatrix}u(1-u)^3 + 6\begin{bmatrix} 2 \\ 4 \\ 0 \end{bmatrix}u^2(1-u)^2 + 4\begin{bmatrix} 3 \\ 3 \\ 0 \end{bmatrix}u^3(1-u) + \begin{bmatrix} 4 \\ 0 \\ 0 \end{bmatrix}u^4$$

$$r_2 = \begin{bmatrix} 0 \\ 0 \\ 4 \end{bmatrix}(1-u)^4 + 4\begin{bmatrix} 0 \\ 0 \\ 4 \end{bmatrix}u(1-u)^3 + 6\begin{bmatrix} 2/3 \\ 2 \\ 4 \end{bmatrix}u^2(1-u)^2 + 4\begin{bmatrix} 2 \\ 6 \\ 4 \end{bmatrix}u^3(1-u) + \begin{bmatrix} 4 \\ 0 \\ 4 \end{bmatrix}u^4$$

Now these are two planar curves, one in the plane z = 0, the other in the plane z = 4. Furthermore they have identical shape: they are both parabolic. In fact they are derived from the same Bézier quadratic form for a segment of a parabola by redefining the parametrization in different ways. The parametrization, it will be recalled, does not influence the shape of the curve. If we apply the above procedure for pro-ducing a surface between these two segments, we certainly obtain a ruled surface, but it is hardly the one we would have hoped to have generated; it is shown in Figure 5.2. (Even more unacceptable results are obtained if the parametrization on one of the curves is reversed.)

Ruled surface patches are useful. But they have limitations in that they are not truly free-form surfaces since they are

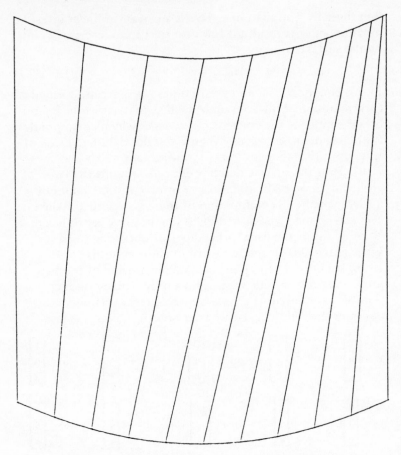

Figure 5.2 *Ruled surface with non-natural point correspondences*

constrained to be linear in one direction. Furthermore
Equation (5.1) relies upon the equations for each of the
defining curves rather than being a generic form.

The next stage in the development is to look at surfaces
that can be defined when four bounding curves are given and
this is the idea of Coons' patches.

Coons' patches
This technique was originated by Coons in the late 1960s and
has subsequently been developed (Coons, 1967; Forrest,
1972). The ruled surface idea apparently demands the
provision of two curves. In a sense however four segments are

used in the definition: the two that are given and two straight lines connecting their corresponding ends. These four form the boundary of the patch. Coons' patches represent an attempt to allow all four of the bounding curves to have general forms.

Suppose that the four given curves are:

$$r_0(u), r_1(u), s_0(v), s_1(v)$$

These define the boundary of the patch to be created. They are shown schematically in Figure 5.3. As usual the parameter

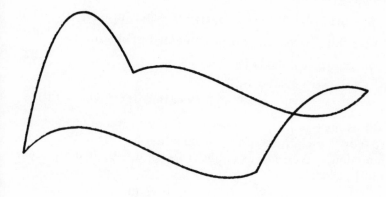

Figure 5.3 *Boundary curve for a patch*

values u and v go between 0 and 1. In order that the boundary joins up correctly we require the following relations to hold:

$$r_0(0) = s_0(0) \qquad r_0(1) = s_1(0)$$
$$r_1(0) = s_0(1) \qquad r_1(1) = s_1(1)$$

If we used the ruled surface approach, we could combine these in two different ways by looking at the pairs of opposite curves:

$$(1 - v)r_0(u) + vr_1(u)$$

$$(1 - u)s_0(v) + us_1(v)$$

If we combine these two surfaces by addition, then along each boundary of the patch we have one of the defining curves which we do want, and a straight line interpolation which we do not want. We attempt to remove the latter by defining the following patch which is formed as a ruled surface (in both directions) with a boundary formed of straight lines joining the four corner points:

$$(1 - u)(1 - v)r_0(0) + u(1 - v)r_0(1) + (1 - u)vr_1(0) + uvr_1(1) \quad (5.2)$$

These three parts can now be combined to obtain:

$$\begin{aligned}
r(u, v) = &(1 - v)r_0(u) + vr_1(u) \quad\quad\quad\quad\quad\quad\quad\quad (5.3) \\
&+ (1 - u)s_0(v) + us_1(v) \\
&- (1 - u)(1 - v)r_0(0) - u(1 - v)r_0(1) - (1 - u)vr_1(0) - uvr_1(1)
\end{aligned}$$

This is the patch equation we want. If we set u or v equal to 0 or 1, then the result is the equation for one of the given bounding curves. For example, if we set $u = 0$, then Equation (5.3) yields:

$$\begin{aligned}
r(0, v) = &(1 - v)r_0(0) + vr_1(0) \\
&+ s_0(v) \\
&- (1 - v)r_0(0) - vr_1(0) \\
= &\, s_0(v)
\end{aligned}$$

Figure 5.4 shows the various stages in the build up of the patch: the originally specified bounding curves; the first two ruled surfaces produced; the surface ruled in both directions; and the combination giving the final form.

It should also be noted that the Coons' patch forms in Equation (5.3) also include the ruled surface form. If one pair of opposite bounding curves are taken to be straight lines, then the equation simply reduces to that of Equation (5.1).

We have used the linear interpolation functions $(1 - u)$ and u in the above derivation. These particular ones are not

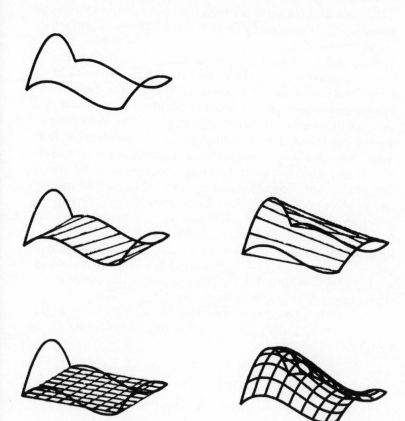

Figure 5.4 *Building up a patch*

essential. If $\lambda_0(t)$ and $\lambda_1(t)$ are functions with the properties that:

$$\lambda_0(0) = 1 \qquad \lambda_0(1) = 0$$
$$\lambda_1(0) = 0 \qquad \lambda_1(1) = 1$$

then a more general form of Equation (5.3) is possible, namely:

$$
\begin{aligned}
\mathbf{r}(u, v) = \; & \lambda_0(v)\mathbf{r}_0(u) + \lambda_1(v)\mathbf{r}_1(u) \\
& + \lambda_0(u)\mathbf{s}_0(v) + \lambda_1(u)\mathbf{r}_1(v) \\
& - \lambda_0(u)\lambda_0(v)\mathbf{r}_0(0) - \lambda_1(u)\lambda_0(v)\mathbf{r}_0(1) \\
& - \lambda_0(u)\lambda_1(v)\mathbf{r}_1(0) - \lambda_1(u)\lambda_1(v)\mathbf{r}_1(1)
\end{aligned}
$$

139

This again defines a patch which has the four given curves as boundaries (as may easily be checked by setting u and v to 0 and to 1).

Such weighting functions λ_0 and λ_1 can also be used, not only to ensure that the edges of a patch coincide with prescribed boundaries, but also that the patch leaves the boundaries in specified directions. Let $r(u, v)$ be a patch. If r_u denotes the partial derivative with respect to u, then this specifies the direction of the tangent to the patch along lines where v is constant. In particular along the bounding curves $r(u, 0)$ and $r(u, 1)$, it gives the tangent direction of the curve. We are more interested here in its effect along the bounding curves $r(0, v)$ and $r(1, v)$. Here it specifies the direction in which the patch starts to leave these curves. In the same way r_v denotes the partial derivative with respect to v and it determines the way in which the patch leaves the boundary curves for which v is 0 or 1. In what follows we also need to use r_{uv} which denotes the first mixed partial derivative of r with respect to u and then to v.

We need to choose weighting functions $\lambda_0(t)$, $\lambda_1(t)$, $\mu_0(t)$ and $\mu_1(t)$ in such a way that they satisfy the following relations:

$$\lambda_0(0) = 1 \qquad \lambda_1(0) = 0$$

$$\lambda_0(1) = 0 \qquad \lambda_1(1) = 1$$

λ_0 and λ_1 have zero derivatives at $t = 0$ and $t = 1$

μ_0 and μ_1 are both zero at $t = 0$ and $t = 1$

$$\mu_0{}'(0) = 1 \qquad \mu_1{}'(0) = 0$$

$$\mu_0{}'(1) = 0 \qquad \mu_1{}'(1) = 1$$

Polynomial functions which satisfy these relations need to be at least cubics. The simplest are as follows:

$$\lambda_0(t) = 1 - 3t^2 + 2t^3 \qquad \lambda_1(t) = 3t^2 - 2t^3$$

$$\mu_0(t) = t - 2t^2 + t^3 \qquad \mu_1(t) = -t^2 + t^3$$

We can now present a relation which gives patch positions $r(u,v)$. These are expressed in terms of the values of this function and its derivatives on the boundary. This is purely for notational convenience. The practical situation occurs when the function r is unknown, but its form and that of its derivatives are known on the four bounding edges. By specifying what the derivatives have to be, we are allowing the possibility of putting distinct patches together in such a way that they join up smoothly. The formula for constructing **r** is as follows. It is written formally as a matrix expression with vector entries:

$$\mathbf{r}(u, v)$$

$$= [\lambda_0(u)\ \lambda_1(u)\ \mu_0(u)\ \mu_1(u)] \begin{bmatrix} \mathbf{r}(0, v) \\ \mathbf{r}(1, v) \\ \mathbf{r}_u(0, v) \\ \mathbf{r}_u(1, v) \end{bmatrix} \tag{5.4}$$

$$+ [\mathbf{r}(u, 0)\ \mathbf{r}(u, 1)\ \mathbf{r}_v(u, 0)\ \mathbf{r}_v(u, 1)] \begin{bmatrix} \lambda_0(v) \\ \lambda_1(v) \\ \mu_0(v) \\ \mu_1(v) \end{bmatrix}$$

$$- [\lambda_0(u)\ \lambda_1(u)\ \mu_0(u)\ \mu_1(u)] \begin{bmatrix} \mathbf{r}(0, 0) & \mathbf{r}(0, 1) & \mathbf{r}_v(0, 0) & \mathbf{r}_v(0, 1) \\ \mathbf{r}(1, 0) & \mathbf{r}(1, 1) & \mathbf{r}_v(1, 0) & \mathbf{r}_v(1, 1) \\ \mathbf{r}(0, 0) & \mathbf{r}(0, 1) & \mathbf{r}_{uv}(0, 0) & \mathbf{r}_{uv}(0, 1) \\ \mathbf{r}(1, 0) & \mathbf{r}(1, 1) & \mathbf{r}_{uv}(1, 0) & \mathbf{r}_{uv}(1, 1) \end{bmatrix} \begin{bmatrix} \lambda_0(v) \\ \lambda_1(v) \\ \mu_0(v) \\ \mu_1(v) \end{bmatrix}$$

Again it can be checked, by taking u and v to be 0 or 1, that this defines a patch with the appropriate values for **r** and its derivatives on the boundary.

Surfaces through given points

In some ways this may seem a step backwards. In the last section we discussed patches that included given curves – a more general problem than putting patches through points. However it is useful to look at the results of the last section in a different way. If we are given four points $\mathbf{a}_{00}$, $\mathbf{a}_{10}$, $\mathbf{a}_{01}$ and $\mathbf{a}_{11}$ in space, then it is easy to put a surface patch through them. This is essentially Equation (5.2) and the patch is:

$$\mathbf{r}(u, v) = (1 - u)(1 - v)\mathbf{a}_{00} + u(1 - v)\mathbf{a}_{10} + (1 - u)v\mathbf{a}_{01} + uv\mathbf{a}_{11}$$

This is a ruled surface in both directions. It is the equivalent for surfaces of forming a curve segment between two points by joining them with a straight line. In Chapter 3, we looked at a method for putting a curve composed of many Bézier segments through a number of points in space. This worked by first establishing for each point an appropriate tangent direction. We can adopt this approach when trying to fit a surface composed of a collection of patches through a number of given points in space. We need to assume that the points are given in such a way that, when we have chosen suitable u and v parameter directions, they may be regarded as lying on a rectangular grid. That is to say, we must be able to label them as $a(i, j)$ for $i = 1, 2, \ldots, m$ and $j = 1, 2, \ldots, n$; the total number of points being mn. The reason for putting the values i and j in brackets rather than as subscripts is purely for convenience as we need to use subscripts to denote derivative values. We now put in four-sided patches, a typical one of which has its corners at the points:

$$a(i, j), \; a(i + 1, j), \; a(i, j + 1), \; a(i + 1, j + 1)$$

By analogy with the case of curve segments, we need to have specified or to have had selected values for the partial derivatives at these corner points. These are assumed known and are denoted by $a_u(i, j)$ and $a_v(i, j)$. Additionally we need to have a value for the first cross-derivative at each point; this is denoted by $a_{uv}(i, j)$. A formula similar to Equation (5.4) at the end of the last section defines the patch we want in terms of the interpolating functions λ_0, λ_1, μ_0 and μ_1; it is given in matrix notation for conciseness:

$$r(u, v) = L(u)^T M L(v) \qquad (5.5)$$

where

$$M = \begin{bmatrix} a(i, j) & a(i, j + 1) & a_v(i, j) & a_v(i, j + 1) \\ a(i + 1, j) & a(i + 1, j + 1) & a_v(i + 1, j) & a_v(i + 1, j + 1) \\ a_u(i, j) & a_u(i, j + 1) & a_{uv}(i, j) & a_{uv}(i, j + 1) \\ a_u(i + 1, j) & a_u(i + 1, j + 1) & a_{uv}(i + 1, j) & a_{uv}(i + 1, j + 1) \end{bmatrix}$$

and

$$L(t) = [\lambda_0(t) \; \lambda_1(t) \; \mu_0(t) \; \mu_1(t)]^T$$

It is easily checked that this patch does indeed go through the required four corner points. Additionally it can be seen that each edge of the patch defined is dependent only upon its end-points and the partial derivative values at the ends in the direction of the edge. These are combined using the interpolation functions λ_0, λ_1, μ_0 and μ_1. The partial derivative giving the tangent direction away from each edge depends upon the values of these at the end-points and upon the values of the mixed derivatives at the end-points; again the precise formula is one involving these and the interpolating functions. The point is that each edge and the tangent direction across it depend only upon values associated with its end-points. Thus these values are the same when that edge is taken to be in the adjacent patch. Consequently we have continuity of the surface and of the tangent direction when all the patches are put together. This gives the required surface through the grid of given points.

As indicated above the patch here described has a form like part of that given by Equation (5.4). That formula dealt with a patch developed when the boundary curves and the derivatives associated with it were already known. If these happen to be of the form described above for the patch produced here, that is depend upon properties of the end-points combined using the interpolating functions, then the three terms in Equation (5.4) become identical, two of them cancel and we are left with Equation (5.5) (Faux and Pratt, 1979).

In the last section, it was seen that the simplest inter-polating functions had to be cubic polynomials. Thus the patch given by Equation (5.5) has edges which are cubic polynomial functions. In Chapter 2 it is seen that we can always write cubic polynomials in Bézier form, and by doing this we can recast the patch here developed as a Bézier cubic patch. This type of patch and its properties are discussed in the next section. Patches like the one in Equation (5.5) and the Bézier ones of the next section have equations which are sums of terms each of which is a product of a function of u and a function of v. These are called 'tensor product' surfaces.

Bézier surface patches

In Chapter 3 Bézier curve segments are defined in terms of the Bézier basis functions. The products of each with a control vector are taken and added together to give a typical point on the curve segment. The same procedure is used here to define the Bézier surface patch except that we need to take products of basis functions and a double sum is used. The form of the patch formed from cubic basis functions is:

$$\mathbf{r}(u, v) = \sum_{i=0}^{3} \sum_{j=0}^{3} \binom{3}{i} \binom{3}{j} a_{ij} u^i (1-u)^{3-i} v^j (1-v)^{3-j} \quad (5.6)$$

(In fact we can obtain the form which is of degree n in both u and v by replacing 3 by n everywhere in the above expression.) As usual u and v are parameters which take values between 0 and 1. The above form, cubic in u and v, contains 16 control points a_{ij}. These determine the shape of the patch. They can be regarded as forming a polyhedron in space and this is illustrated in Figure 5.5.

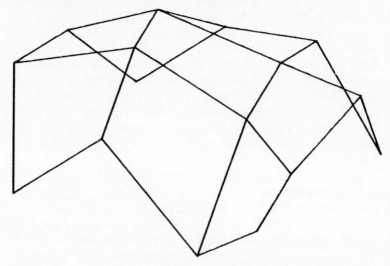

Figure 5.5 *Polyhedron of patch control points*

If we set u and v both equal to zero, then we find that the value of $r(0, 0)$ is simply a_{00}; as usual we need to take zero raised to the zeroth power to be unity. Thus this control point lies on the patch. Similarly, the corner points of the control polyhedron a_{03}, a_{30}, a_{33} together with a_{00} all lie on the patch and form the corners of it. These are the only control points which in general lie on the curve.

If we set $v = 0$, then we find that:

$$r(u, 0) = \sum_{i=0}^{3} a_{10}u^i (1 - u)^{3-i}$$

This is precisely the form of a Bézier curve segment whose control points are four points from one side of the control polyhedron for the patch. In this way the edges of the patch are Bézier curves determined by the control points along the edges of the polyhedron. This can be seen in Figure 5.5 which shows the patch together with its control polyhedron. Thus 12 out of the 16 control points are associated with the edges of the patch; the other four help to influence the shape internally.

Bézier surfaces are implemented on many commercial CAD system. Their popularity is due to the comparatively simple form of description which is general enough for most applications and yet depends on a small number of control points which can be stored in the node list. They can be used to represent many standard surface forms. For example, the ruled surfaces and Coons' patches discussed earlier can both be reformulated in Bézier form. Many of the properties of the Bézier curve segment also apply to the surface patches. For example, the patch lies within the convex hull of the control points. We can transform the entire surface patch by applying the transformation to each of the control nodes individually.

Displaying surfaces on a graphics output device is something of a problem. It is usually solved by displaying 'isoparametric curves'. These are simply the curves in the patch produced by keeping one of the parameters fixed and letting the other vary between 0 and 1. By doing this for both parameters, we can build up a grid on the surface which is usually sufficient to permit visualization of the surface. This technique is used in Figure 5.6. Commercial CAD systems normally provide a command for specifying how many isoparametric curves in each direction are to be used. This grid can also be used for other purposes. For example if the grid is made reasonably fine, the quadrilaterals formed on the surface have curved sides which are close enough to straight line segments to be taken as such. The area of each part can then be evaluated and the area of the entire patch estimated. For more sophisticated analysis, this way of obtaining a grid can be used as a starting point for the automatic generation of finite element meshes covering the surfaces of the component being designed. The subdivision process is also the basis of facetting techniques which can be used to speed display procedures and permit shading of surfaces. We discuss these in more detail in Chapter 7.

Bézier patches can be defined for any degree polynomials; the degrees in u and in v can be different if required. Rational surface forms are also possible by introducing homogeneous coordinate vectors as control points into Equation (5.6). We end this section with some examples of these. For convenience the degree is reduced to 2. The restriction this

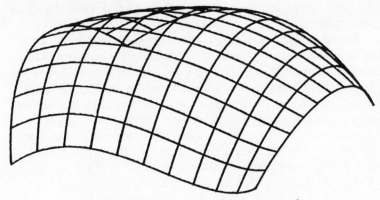

Figure 5.6 *Isoparametric curves on a patch*

imposes is to some extent counterbalanced by the extra flexibility provided by the additional coordinate; however the patches produced are incapable of having saddle points and other inflexions.

Patches of degree 2 in u and v have nine control points in their definition; of these eight control the bounding edges of the patch. So if we already know what the boundary is we have only to select the ninth control point to complete the definition of the surface. As an example consider the octant of a torus shown in Figure 5.7. The bounding curves are circular arcs and so can be represented exactly by rational

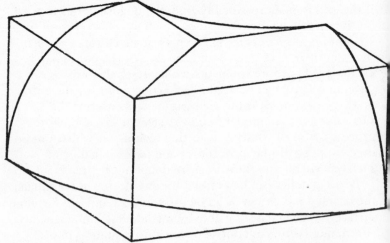

Figure 5.7 *Octant of a toroidal patch with control polyhedron*

quadratics. Four of the patch control points come from the corners of the patch, four more from the intersections of the end-tangents of the bounding curves (with the homogeneous weighting chosen as described in Chapter 3). A little experimentation is needed to find the correct ninth control point but once this is done the Bézier patch exactly produces the torus octant. The actual control points are as follows and a representation of the surface using isoparametric curves is given in Figure 5.8:

$$\mathbf{a_{00}} = [1 \quad 0 \quad 1 \quad 1]^T$$
$$\mathbf{a_{01}} = [1/\sqrt{2} \quad 1/\sqrt{2} \quad 1/\sqrt{2} \quad 1/\sqrt{2}]^T$$
$$\mathbf{a_{02}} = [0 \quad 1 \quad 1 \quad 1]^T$$
$$\mathbf{a_{10}} = [\sqrt{2} \quad 0 \quad 1/\sqrt{2} \quad 1/\sqrt{2}]^T$$
$$\mathbf{a_{11}} = [1 \quad 1 \quad 1/2 \quad 1/2]^T$$
$$\mathbf{a_{12}} = [0 \quad \sqrt{2} \quad 1/\sqrt{2} \quad 1/\sqrt{2}]^T$$
$$\mathbf{a_{20}} = [2 \quad 0 \quad 0 \quad 1]^T$$
$$\mathbf{a_{21}} = [\sqrt{2} \quad \sqrt{2} \quad 0 \quad 1/\sqrt{2}]^T$$
$$\mathbf{a_{22}} = [0 \quad 2 \quad 0 \quad 1]^T$$

Figure 5.9 shows a truncated cone shape made up of 12 rational quadratic patches. It also shows the positions of these patches. The ones forming the edge radius are again toroidal and have a form like that above. The ones which produce the sloping curved sides are produced similarly; in fact they are ruled surfaces and this provides an example of a

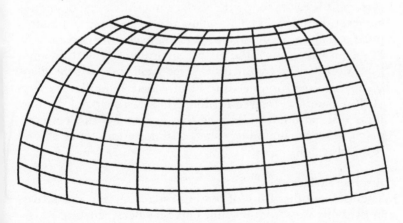

Figure 5.8 *Toroidal quadrant with surface patch filled in*

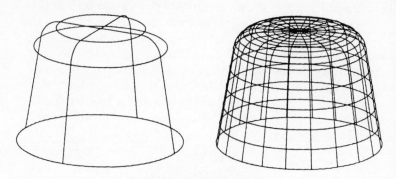

Figure 5.9 *Truncated cone*

ruled surface cast in Bézier form. Finally the patches on the flat top of the cone seem to have a triangular rather than a quadrilateral boundary. We have cheated slightly. For each of these patches at the point which corresponds to the centre of the top are defined three coincident control points; this means that the missing fourth side is one of zero length positioned at this centre point. The form of the isoparametric curves shows that this is the case.

This cone example shows that surface patches can be put together to produce everyday shapes. In general, however, care has to be taken to ensure that the patches fit together smoothly; the problems involved in this are discussed later in this chapter. From the point of view of a user of a CAD system, it is easier to input a surface in terms of its bounding curves. Commercial systems allow this to be done and inter-polate an appropriate surface patch between the curves. Problems can occur if the surface so produced is not the one that was intended by the user. Some facility to edit the shape to the required form needs to be provided. In Mullineux (1984) a method is given for filling in a surface patch in this type of situation; it is designed that it produces the 'correct' patch when the bounding curves outline certain types of toroidal region.

B-spline surfaces

These are another form of tensor product surface and derive in the same way from B-spline curves as Bézier surface patches do from Bézier curves. Indeed the form of equation is identical. The most general is:

$$\mathbf{r}(u, v) = \sum_{i=0}^{m} \sum_{j=0}^{n} a_{ij} p_i(u) q_j(v) \qquad (5.7)$$

Here the p_i and the q_j are B-spline basis functions as defined in Chapter 3. They are usually different sets of functions as we can choose different sets of knots for u and for v.

As with Bézier patches, we have a number of control points which can be regarded as forming a polyhedron which influences the shape of the surface. The advantage with the B-spline form is that adjusting the position of a control point only affects the surface locally. If we talk about a span of the surface as being that part generated as u and v move between two consecutive knot values for each, then any control point affects only 16 spans of the surface if cubic basis functions are being used. Conversely, any span is generated by just 16 of the control points and it lies within the convex hull of these points. Thus the surface tends to follow the same sort of shape as does the defining polyhedron.

Some CAD systems provide access to B-spline surfaces. It needs to be borne in mind that they are powerful tools with a great deal of flexibility inherent in them. As a consequence they are difficult to control. CAD systems tend to hide some of this flexibility away so that the user is not required to deal with the surface definitions directly; they provide semi-automatic procedures for manipulation, for example by inserting B-spline surfaces through given points or to fill in specified gaps between existing geometric entities. However it is difficult to ensure that such procedures are completely foolproof. Peculiarities can occur and it is wise always to check the form of any surface generated to ensure it is what is required. As an example of the problems that can occur, Figure 5.10 (obtained from Hammond, 1983) shows the results of attempting to fit a B-spline surface over a bottle shape. Very unexpected results were obtained including a large number of cusps which were certainly not required.

The DUCT system

DUCT is a commercially available CAD system designed originally as a means of producing surfaces in a way that could be easily understood by a user. As its name suggests it is based upon producing basic shapes which are like the ducting on car manifolds and indeed this was one early application

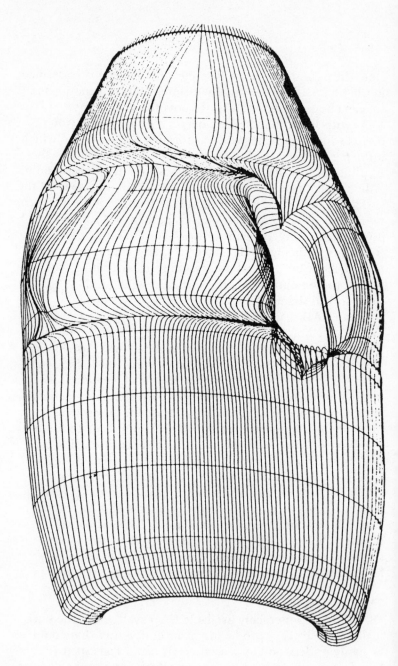

Figure 5.10 *Erroneous surface for a bottle shape*

of the system. It is also being used in the production of moulds and dies.

The basic construction is to form a surface which connects together a number of predefined curves which could be open or closed. In this respect it resembles the technique of 'lofting'. This is used in the aircraft and ship building industries to enable the definition of large smooth surfaces. It applies to surfaces, such as aircraft wings, which are considerably longer in one direction than the other. Sections across the surface normal to the long axis are drawn out and then these are smoothed together to define the final shape. In some applications where the precise shape is not critical, this smoothing operation may be left to the pattern or tool maker who is just provided with the details of the sections.

In DUCT, a number of surface entities called ducts are defined interactively by the user. Each has a long axis which is defined by the user as being a smooth curve called the 'spine'. This need not be a straight line and it is possible to allow it to twist about quite considerably. About the spine, the user defines a number of curves; these may be open curve segments or they can be closed. These are the cross-section curves. They are planar and the system ensures that each plane is normal to the spine curve at the point where it cuts it. There is no need for the section curves to have the same shape, although in practice a user would ensure that there was only gradual change from one to the next.

By smoothly joining together corresponding points on consecutive section curves (cf. the formation of smooth curves through a number of points discussed in Chapter 3), a number of four-sided regions in space are created and these can be smoothed over with surface patches to complete the definition of the duct. An example of a duct defined in this way is shown in Figure 5.11.

When several ducts have been created, they can be combined together to produce a surface description of the required component. A wide variety of component shapes can be produced in this way and the system provides a reasonably simple way of defining complex surface geometry. Some examples of components designed using DUCT are shown in Figures 5.12 and 5.13.

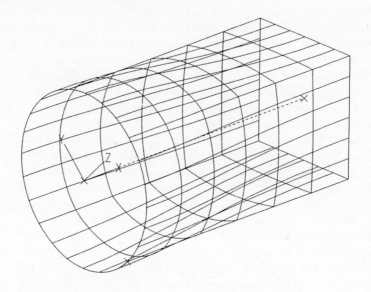

Figure 5.11 *Simple duct (courtesy of Deltacam Systems Limited)*

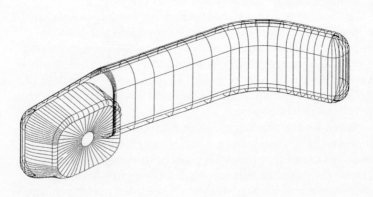

Figure 5.12 *Telephone handset (courtesy of Deltacam Systems Limited)*

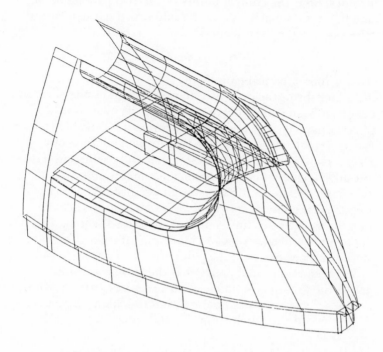

Figure 5.13 *Domestic iron (courtesy of Deltacam Systems Limited)*

Problems involved in putting patches together

We concentrate in this section upon Bézier cubic patches since they are fairly straightforward and demonstrate well the ideas involved. As might be expected, many of the ideas do extend and apply to more sophisticated Bézier and B-spline forms. We are interested in ensuring that if a large surface area is being designed using a large number of individual patches, then these go together in such a way that a smooth overall result is achieved.

To obtain this smoothness we certainly need to arrange that the patch boundaries meet. The boundary of a Bézier cubic patch is a Bézier cubic curve segment controlled by the four points along the relevant side of the polyhedron. Consequently the easiest way of ensuring that patches come together correctly is to use the same control points for the two touching boundaries of the patches. We are assuming that the patches are to join along the entire length of a pair of sides. So if the control points of the two patches are a_{ij} and b_{ij} and they meet along the side which has $u = 1$ on the first and $u = 0$ on the second, then we require that:

$$a_{3j} = b_{0j} \qquad \text{for } j = 0, 1, 2, 3 \tag{5.8}$$

This is illustrated in Figure 5.14.

In fact this is the only way to proceed for cubics since in general a Bézier cubic curve determines its own control points uniquely. For higher order forms it is possible to have two different sets of control points producing the same shape. However the above approach is still useful and has the advantage that corresponding parameter points on the two curves relate to the same point on space. Thus if an isoparametric line is drawn on each patch up to the common boundary, then this line is unbroken across the join. This has application when it comes to machining the entire surface since this need not be done patch by patch.

We need also to make sure that the surface is smooth across the join of two patches. This requires us to take into account tangency considerations and to look at the partial derivatives of the patch form. for the patch:

$$r(u, v) = \sum_{i=0}^{3} \sum_{j=0}^{3} a_{ij} u^i (1-u)^{3-i} v^j (1-v)^{3-j}$$

the partial derivatives with respect to u and to v are given by:

$$r_u(u, v) = 3 \sum_{i=0}^{2} \sum_{j=0}^{3} (a_{i+1,j} - a_{i,j}) u^i (1-u)^{2-i} v^j (1-v)^{3-j}$$

$$r_v(u, v) = 3 \sum_{i=0}^{3} \sum_{j=0}^{2} (a_{i,j+1} - a_{i,j}) u^i (1-u)^{2-i} v^j (1-v)^{3-j}$$

Now the direction of the normal to the surface at any point is given by the vector product of these two derivatives:

$$\text{surface normal} = r_u \wedge r_v \tag{5.9}$$

To ensure smoothness of the surface we need to have smoothness of the tangent plane and this is equivalent to continuity of the direction of this surface normal vector.

Before pursuing this further we look at the above partial derivatives in terms of what happens at the corners of the patch as we did when discussing Coons' patches earlier in this

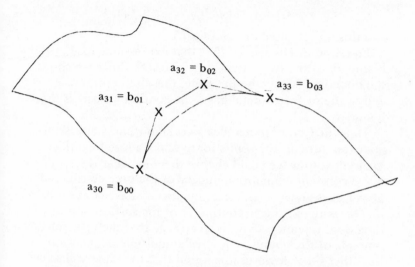

Figure 5.14 *Control points at the junction of two patches*

chapter. We find that the various expressions simplify considerably and we obtain the following for the corner where $u = v = 0$:

$$\mathbf{r}(0, 0) = \mathbf{a}_{00}$$
$$\mathbf{r}_u(0, 0) = 3(\mathbf{a}_{10} - \mathbf{a}_{00})$$
$$\mathbf{r}_u(0, 0) = 3(\mathbf{a}_{01} - \mathbf{a}_{00})$$
$$\mathbf{r}_{uv}(0, 0) = 9(\mathbf{a}_{11} - \mathbf{a}_{01} - \mathbf{a}_{10} + \mathbf{a}_{00})$$

Similar relations hold for the other corners. Now the philosophy behind Equation (5.4) for example is that the value of these vector quantities should be specified at points on a grid in space and used to define a surface. If this is done, then the above equations applied at four points which define the corners of a patch are sufficient to specify all 16 control points precisely. Furthermore if we use these vectors to obtain the control points $\mathbf{b}_{ij}$ of the neighbouring patch (with the $v = 1$ edge of the first lying along the $v = 0$ edge of the second as above) we find that in addition to Equation (5.8) we have:

$$\mathbf{a}_{3j} - \mathbf{a}_{2j} = \mathbf{b}_{ij} - \mathbf{b}_{0j} \qquad j = 0, 1, 2, 3. \qquad (5.10)$$

This says that across the boundary we have sets of collinear control points; these points being:

155

$$a_{2j}, a_{3j} = b_{0j}, b_{1j} \qquad j = 0, 1, 2, 3$$

and this is illustrated in Figure 5.15.

Equation (5.10) also ensures that expressions for r_u on the boundary of the two patches are identical. Since we choose the bounding edge to be the same, this also applies to r_u and so the surface normal has to be continuous across the boundary.

Thus the Coons' patch idea gives smoothness but determines all our control points for us and we lose flexibility. We look at how we could choose them in other ways. Along a v = constant common boundary of the type considered above, as we have seen, we automatically have continuity of r_v. One way to achieve continuity of the surface normal direction, Equation (5.6), is to arrange that the value of r_u on one side of the boundary is a scalar multiple of the value on the other side. However it is found that the same scalar has to be used all along the edge of the patch and indeed has to be used where the common boundary curve continues for the next and subsequent patches. This does not provide much more flexibility than does Equation (5.10) (where a scalar multiple of unity has been used).

We have been trying to ensure continuity for the surface normal direction by looking for continuity in direction of the

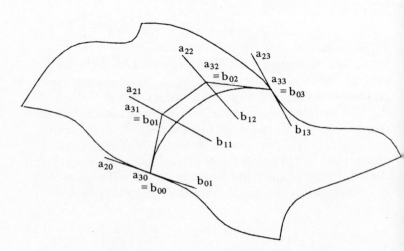

Figure 5.15 *Control points preserving tangent continuity*

two parts involved in the vector product (5.6). While this is a sufficient method of approach, it is not necessary; more general ways are discussed in Faux and Pratt (1979) for example.

What is clear is that the need to achieve smoothness severely limits the flexibility in the choice of control points. One way of overcoming this for Bézier patches is to use ones of higher degree. For example if we use quintic patches there are 25 control points at our disposal. We might attempt to fit a smooth cubic surface first. Any cubic polynomial is also a quintic polynomial (with its two leading coefficients set to zero) and we can reformulate a cubic patch as a quintic one. If we do not want to disturb the smoothness at the boundaries produced by the cubic version, then of the 25 control points established we may not adjust those on the boundary of the polyhedron or those one step in from the boundary. This accounts for 24 points and we are reduced to using the one remaining in the middle of the polyhedron to do any fine tuning of the shape.

MODCON: an example system

Background to the system

MODCON is a system developed in the late 1970s mainly for the computer-aided manufacture (CAM) (rather than design) of dies for the hot forging industry (Chan and Knight, 1979, 1980; Chan *et al.*, 1979; Mullineux and Knight, 1980). Its name is a shortened form of MODular CONstruction; the basic concept is the building up of component shapes out of simpler volumetric forms or primitives. These primitive shapes include such entities as (truncated) cones and rectangular blocks. The general forms of these are predetermined but they possess parameters which the user can define to create the types of shape that are required.

Within the forging industry the components produced are highly varied. The types of the die cavity required are certainly three-dimensional but are, in a larger number of cases, relatively simple in their form. As a consequence they lend themselves to being broken down into simple building blocks which are common geometric entities, and the same shapes are repeated frequently; thus the number of primitive shapes needed is comparatively small. Additionally, it is natural to think of the components in terms of these simpler sub-entities, so the breakdown process is not a difficult one. Very often forging designs do not specifically indicate the transition surfaces to be introduced between the main features of a component. These are needed to ensure satisfactory flow of metal during the forging process. Very often they are simply generated or created when the die is made as a consequence of the method used for manufacture.

The traditional method of die manufacture involves the production of the cavity by direct means. This is a time-consuming procedure requiring a high degree of skill in the personnel involved. More recently copy-milling and electro-discharge manufacture (EDM) have begun to be used. These have considerable advantages when it comes to resinking or remaking dies that have become damaged during operation. This is because the pattern for the die can be made in more easily worked materials than tool-steel and it can be reused or recreated easily.

Manual methods for pattern making also tend to be highly skilled and time-consuming. The process can be facilitated by the use of NC machine tools provided the means to generate the NC tool path is available. It was to this end that MODCON was developed. It was aimed primarily at the pro-duction of EDM electrodes for the spark erosion of finishing dies.

The reason for discussing MODCON in some detail here is that it provides an application for many of the topics intro-duced in previous chapters. It is a fairly simple system to understand the workings of and yet it is powerful enough to be able to generate usable models of real-world components.

As indicated above, it is essentially a system for CAM rather than CAD. It does not have the capabilities required to perform a great amount of design manipulation; it assumes that the user knows where the primitive shapes are to go and processes them accordingly. Nonetheless it illustrates many of the ideas involved in more recent three-dimensional commercial CADCAM systems.

The use of primitive shapes

In order to make use of the MODCON system, it is necessary for the user to describe his component as being built up of a collection of standard (predefined) primitive shapes. These shapes include such items as (truncated) cones, horizontal cylinders, and toroidal elements (see Figure 6.1). Each of these is defined within the system in terms of a number of parameters whose values the user can assign to determine the precise size and position of the primitive element. To illus-trate the types of command available for defining a primitive shape, we consider two examples: the cone and the cylinder primitives.

The cone primitive is shown in Figure 6.2; it is a trun-cated cone. The form of the basic cone is defined by five parameters: the upper radius (UR), the lower radius (LR),

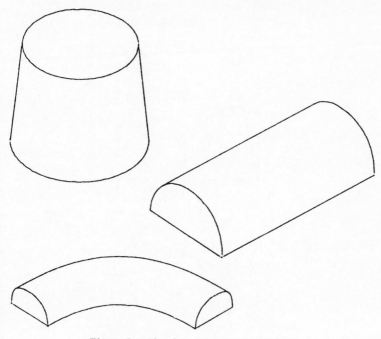

Figure 6.1 *Simple solid primitive shapes*

the edge radius (ER), the vertical height (VH) and the draft angle (DA). In fact these parameters over-define the form, and any four out of the five should be supplied by the user. These then define the basic cone shape. It can be positioned on the plane by specifying values for OX and OY the coordinates of the position of the centre of the cone. Positioning is only allowed in two dimensions by the system since this is sufficient for a large number of forging. The plane upon which the primitives sit is in effect the plane upon which the two halves of the die meet.

To allow a greater number of forms to be derived from the single cone primitive, it is possible to deform it by stretching it in both the x and y directions. The value of XSEP specifies the separation in the x direction and creates a plan view as shown in Figure 6.3. The values of YSEPL and YSEPR specify different y separation values for the left and the right parts of the primitive. These are shown in Figure 6.4. These separation values destroy the axi-symmetry of the cone primitive, so the user is allowed also to rotate the shape

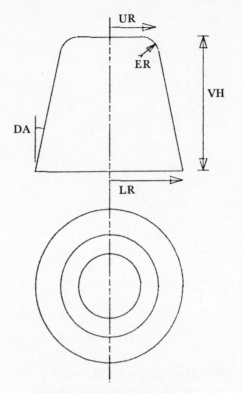

Figure 6.2 *A cone primitive*

about a vertical axis by specifying an angle ROT (in degrees) through which it should be turned (in an anticlockwise direction).

The form of user command that is necessary to define a cone primitive is the command name CONE, followed by the number of the cone (to distinguish one cone from another), followed in turn by values of the various parameters. Thus one possible command to define cone number 3 would be:

CONE 3, VH=50.0, DA=10.0, UR=15.0, ER=0.5, OX=25.5, XSEP=12.0

The units for linear dimensions are given in millimetres and angular ones are in degrees. If a parameter is omitted it is taken to be zero (unless it can be deduced from the other information given).

The cylinder primitive is somewhat less complicated as it

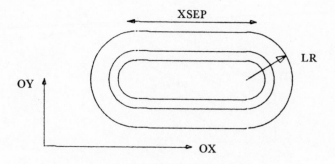

Figure 6.3 *Deformation of a cone primitive*

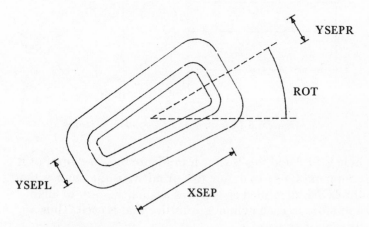

Figure 6.4 *More severe deformation of a cone primitive*

needs fewer parameters to specify it. The basic shape is half a
cylinder lying on the horizontal plane. Its length is specified
by a parameter L, the radius by R, and the coordinates of
its centre by OX and OY. These are shown in Figure 6.5.
Additionally ROT specifies an anticlockwise rotation about a
vertical axis through its centre. Draft angles can be given to
the end faces of the primitive by use of the parameters DAL
and DAR; these refer to the left and right end faces before
any rotation is applied. Finally it is possible, as with the
cone, to introduce a separation into the primitive to create a
flat top surface to the horizontal cylinder. The parameter

163

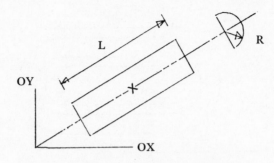

Figure 6.5 *A cylinder primitive*

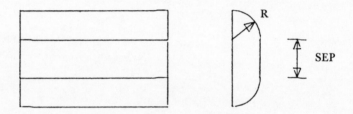

Figure 6.6 *Deformation of a cylinder primitive*

here is SEP and this is shown in Figure 6.6. The form of the command CYLI to define this primitive is the same as that for CONE described above. It is again necessary to give a number to each cylinder primitive that is used. Thus a possible command is:

CYLI 2, R=20.0, L=52.3, DAL=10.0, DAR=10.0, OY=3.5

In addition to solid primitives of the type described so far, it is possible to define ones to represent holes within the component, or more generally regions where material does not exist. The most commonly used of these is the 'plug' primitive. This is essentially an inverted cone and can be manipulated in a similar sort of way. An example of a plug is shown in Figure 6.7.

A further type of primitive is also included within the system; this is the 'dummy'; it is generated by the command DUMY. By employing this, the user can indicate a cuboidal region into which the NC cutter must not enter during

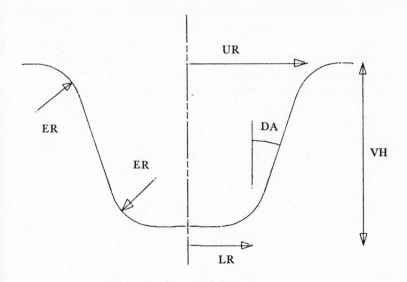

Figure 6.7 *Parameters for a plug primitive*

manufacture of the electrode. This is a simple way, for example, of ensuring that the final tool path generates just one half of a symmetric component; the dummy is placed with one face along the axis of symmetry and completely containing one half of the component. If the NC machine tool is being driven using paper tape, the amount required is reduced and the mirroring capability of the machine tool can be used to generate the other half of the component.

Putting primitives together

The user begins by defining all the primitive shapes that are to go together to make up the component. It is then necessary to fit these together so that parts where the shapes overlap are not machined into when the component is produced. This is achieved by use of the MERGE command. This provides the system with a list of primitives that are to be put together. Additionally up to five values relating to the tool to be used and the finish required are specified. These parameters are supplied using the PARA command which we now discuss.

The first two of these parameter values are T1 and T2 which specify the size of the tool to be used when the electrode is produced by NC milling. These are respectively the radius of the body of the tool and the radius at the end as shown in Figure 6.8. By taking T2 to be zero, the user can specify an end-mill tool and by setting T2 equal to T1 a ball-nosed cutter is identified. Remember that the system is limited to producing output to control a 2.5-axis machine tool.

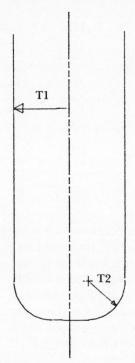

Figure 6.8 *Tool dimensions*

If the values of T1 and T2 are specified too large, then the cutter does not touch the final surface of the component. This can be used to allow the generation of NC commands to produce a roughing cut to remove quickly the majority of stock from the original workpiece. For purposes of displaying the component on a graphics screen before it is machined, the values of T1 and T2 can be set to zero. The contours thus generated by the system and displayed are then curves on the surface of the component.

The finish required on the electrode is determined by the values the user assigns to the parameters EPS and EPSEDG

which refer to the tolerances allowable on deviations within
the machined surfaces. As the tool cuts two consecutive
horizontal contours around the component, it leaves small
cusps. The value of EPS is the maximum height permitted for
those on the main sloping faces of the primitives (Figure 6.9)
and EPSEDG is the value allowed when the various edge
radii on the primitives are produced. The smaller these values

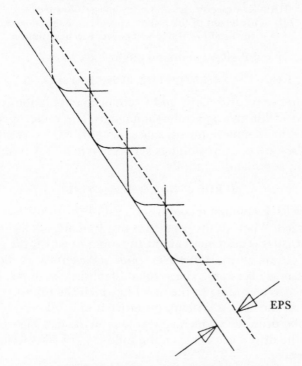

Figure 6.9 *Height of cusps between tool passes*

are set, the better the quality of the finish, but this requires
the vertical steps between successive contours to be smaller
with consequent increases in machining time and in lengths
of paper tape (unless the machine tool can be coupled
directly to the computer). We can influence directly the
maximum vertical distance between contours by giving a
value to the ZIMAX parameter. This specifies the maximum
allowable step between contours and is used if EPS and
EPSEDG would generate a larger value. It is useful to specify

ZIMAX in the case when the sides of a component are nearly vertical or when EPS and EPSEDG have deliberately been set large for testing purposes to check that a component's geometry has been described correctly.

There are other parameters that can be used in a PARA command. These include:

SCALE — scale to be used on the plot of the component
FACTOR — scale factor for the component itself
SHRINK — specifies an overall shrinkage allowance
ZHI — the height of the highest contour to be produced
ZLO — the height of the lowest contour to be produced

Thus a typical PARA command might look like:

PARA TI=25.4/4.0, T2=T1/2, EPS=0.1, EPSEDG=0.1

Note that some arithmetic and algebraic manipulation is allowed within this and other commands. The values specified are used in conjunction with subsequent MERGE commands until they are reset by another PARA command. A typical MERGE command is as follows:

MERGE CONE1, CONE2, CYLI1

The MERGE command comprises a list of the primitives to be merged. When all the primitives involved are solid, there is no particular requirement about the order in which the primitives are given. If however solids and cavities are merged some care is necessary. The system establishes an order of precedence according to the order in which the primitives appear in the list. Any cavity primitive is only allowed to cut into solid primitives which appear later in the list. Figure 6.10 shows the difference between the following two MERGE commands:

MERGE CONE1, CONE2, PLUG1, CONE3
MERGE CONE1, PLUG1, CONE2, CONE3

Only in the second case does the plug enter into cone number 2.

It is possible to use more than one MERGE command to generate tool paths to create the final component. For example, it might be desirable to machine the outside using a large tool to give a good surface finish and to use a smaller tool to remove the material from a hole within the component. In this case the first MERGE command would deal with the solid primitives and a second would specify the

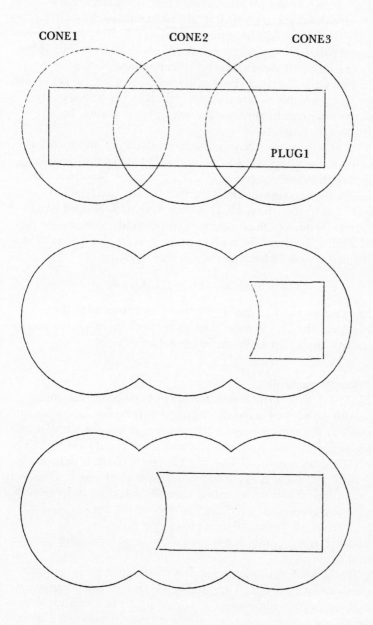

Figure 6.10 *Order of precedence when merging*

smaller tool and maybe just one plug primitive. (This is a special case of the MERGE command; normally cavity primitives which are not part of solid components within the merging list would not generate tool paths. However if only cavity primitives appear the full merged tool paths for them is created.) If it were required to change the tool when dealing with different parts of the outside of the component, say, it is possible to use dummy primitives to guard parts of the component that are known to have already been produced completely.

As the tools used have non-zero radii, they impart fillet radii to the component at points where two primitives apparently come together at a sharp corner. Normally this degree of smoothing is quite sufficient for the types of application for which MODCON is designed. Should larger radii be required, they can be incorporated by means of the BLEND command. This takes as arguments the names of two primitives and a blending radius, for example:

BLEND CONE6, CYLI4, 20.0

The system then examines the generated tool path and wherever the two named primitives meet it adds in a radius to remove the sudden change in tool direction.

A simple example

Table 6.1 shows the commands necessary to define and produce a plot of a simple connecting rod. The exclamation mark at the beginning of a line is used to denote a comment. Six primitives are used in total and these are shown in Figure 6.11. Cones 1 and 2 define the big end; cone 2 is deformed greatly so that it is more or less a rectangular block. Cone 4 is undeformed and represents the small end; cone 3 defines the tapering connector part. The cavities in the big end and the connector are obtained using two plug primitives. As it is defined, the second plug overlaps the region occupied by the first two cones.

The PARA command used has its parameters set so that the contours produced are on the surface of the component and are fairly widely spaced. The plot of the component produced by this sequence of commands is shown in Figure 6.12. Note that the order of the primitives when they are

Table 6.1

```
!   EXAMPLE – A SIMPLE CONNECTING ROD
```

```
!   CONES 1 AND 2 COMPRISE THE BIG END
!   CONE 4 IS THE SMALL END
!   AND CONE 3 IS THE CONNECTING PART

CONE 1, DA=5, UR=16, VH=27, ER=1, XSEP=8
CONE 2, DA=5, UR=1, VH=20, ER=1, XSEP=30, YLSEP=46, YRSEP=46
CONE 3, DA=5, UR=6, VH=16, ER=1, OX=−50, XSEP=80, YRSEP=10
CONE 4, DA=5, UR=8, VH=25, ER=1, OX=−95

!   THE CAVITY IN THE BIG END IS PLUG 1
!   AND PLUG 2 IS THE CAVITY IN THE CONNECTING PART

PLUG 1, UR=8, VHU=27, VHL=12, DA=5, ER=1, XSEP=8
PLUG 2, UR=3.75, VHU=16, VHL=12, DA=5, ER=1, OX=−50, XSEP=80,
                                                          YRSEP=10

!   NOW START TO COMBINE THESE PRIMITIVES

START

!   THE TOOL SIZE IS SET ZERO FOR A PLOT
!   THE PLOTTING SCALE FACTOR IS SET TO FULL SIZE

PARA SCALE=1.0
PARA T1=0, T2=0, EPS=20, EPSEDG=20, ZIMAX=50

!   THE MERGING COMMANDS WITH THE APPROPRIATE
!   PRECEDENCES

MERGE PLUG1, CONE1, CONE2, CONE4, PLUG4, CONE 3

FINISH
```

specified to the MERGE command is such that the cavity in the connector does not intrude into the big end.

In order to produce the electrode to sink the cavity for the component it would of course be necessary to put suitable values into the PARA command to specify the tool to be used and the finish required. Figure 6.13 shows the results of machining the part (in wood rather than graphite).

Operation of the system

When the user specifies the parameters for a primitive shape, the system has sufficient information to determine, for any

171

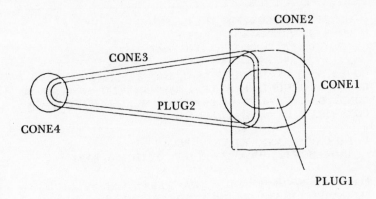

Figure 6.11 *Primitives for the connecting rod example*

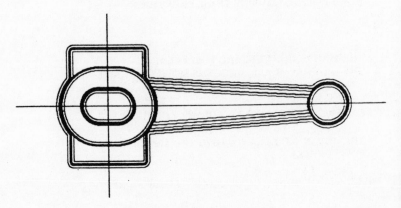

Figure 6.12 *Command sequence*

height z above the plane upon which the primitives sit, a horizontal contour on the surface of the primitive. Furthermore, if the values T1 and T2 for the cutting tool are known, then it is possible to determine the path that the tool must follow at the given height in order just to touch the primitive. (Note that in general the contour in which the tool touches the part will not be at exactly the same height.) If the height above the defining plane is too great, then the system can tell that no contour for the primitive is possible.

When the system encounters a MERGE command, it

Figure 6.13 *Finished model of the component*

proceeds to produce tool path contours at various heights.
For any height, it first produces contours for each primitive
specified in the list associated with the MERGE command.
Because of the nature of the primitive defined in MODCOM,
these contours are fairly simple in shape being composed of
straight lines and circular arcs. They can be held within the
system by means of node and edge lists as discussed in
Chapter 2; each edge is defined by at most three nodes. When
all the contours at the required height have been produced,
it is necessary to try to combine them. This involves intersecting them in pairs. Since the analytical forms of the edges
are simple this can be carried out by solving simple simultaneous equations. If more complicated shapes were allowed,
then methods like those described in Chapter 4 would be
necessary. When the intersections have been found, the next
stage is to reassemble the various parts of the contours so
that the proper outer boundary is obtained. The method

discussed at the end of Chapter 4 can be used here. It is how-ever necessary to ensure that the correct order of precedence between the primitives is respected so that cavities do not intrude wrongly into solid parts of the component.

Once the combined contour has been produced, it repre-sents a path that the cutting tool has to follow. It is then reasonably simple to translate the information into the appropriate form for the machine tool controller.

When a contour at a particular height has been produced, the next height has to be determined. Each primitive in the list of the MERGE command is examined. The next appro-priate height for each is found by using the current EPS or EPSEDG value depending on whether the current cut is on the main sloping side or on the edge radius. The values obtained from each primitive are examined and the one that represents the least step is used. When a primitive has a flat top surface special care is taken. Contours for this surface are produced working outwards from the centre. The height of the cutting tool is not incremented until all the contours necessary to produce the top surface have been produced and have subsequently been intersected with all the other primitives within the MERGE list. The system continues to produce contours in this way until the defining plane upon which the primitives lie is reached (or until the height specified by the ZLO parameter is attained).

As originally developed, MODCON does not allow much facility for controlling the actions of the machine tool apart from its basic movements. It is possible to specify spindle speeds and feed rates through additional parameters to the PARA command, although this is not discussed here. Further control of the cutting operation can be gained by the crude means of writing the NC instructions to the file and then editing the file after the MODCON session.

Limitations of the system

There are two main areas of limitation for MODCON, although both have the advantage that the system itself is small and can run easily on small computers. The first is its reliance on the fact that 2.5-axis machining is to be used in the production of the components. Thus the system need only produce horizontal contours at particular heights. This means that the various primitive shapes are not truly merged

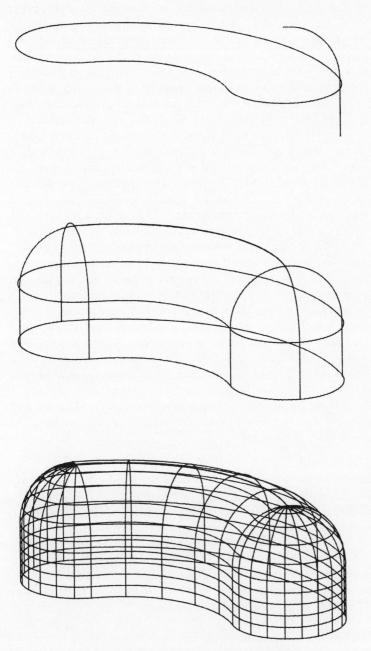

Figure 6.14 *Surface produced by sweeping*

within the system and no complete computer model is generated. Thus the user cannot view the complete component
except as a series of plane contours. While this is sufficient
to allow for visual checking of the geometry, it does require
this geometry to be known in advance. This limits the use of
the system as a design tool as opposed to a manufacturing aid.

The second limitation is the reliance on primitive shapes
that are built into the system. While the ones defined are
quite sufficient for many components, there are occasions
when a shape is difficult to produce from the given primitives.
The ability for a user to add extra primitives for himself
would be an advantage. Approaches by which this could be
done are discussed in Chapter 7 when we look at more
sophisticated modelling systems. One possible approach,
which would be simple for a user to handle, would be to
allow the definition of a curve to represent the profile of a
primitive which would then effectively be spun about a
vertical axis to define an axi-symmetric solid such as the
basic cone primitive. An alternative is again to define a curve
to represent a vertical section through the shape and second
curve which is treated as horizontal and around which the
first curve is moved. An example of this approach is shown in
Figure 6.14. This is essentially the idea of the DUCT system
discussed in the last chapter. Indeed it would be possible to
define all the primitives within the system in this way and
hold them by means of their surface geometry. Here we are
moving into the area of surface modellers which are described
more fully in Chapter 7.

Conclusions

MODCON is a fairly simple system for producing dies for the
forging industry. It is a CAM rather than a CAD system. It
makes use of the fact that a large number of the components
to be produced are composed of simple volumetric shapes
which can be predefined in the system. By limiting itself to
2.5-axis machining, MODCON reduces a three-dimensional
situation to a series of two-dimensional cases by looking at a
set of horizontal contours which represent tool path motions.
These are straightforward to combine.

However, MODCON well illustrates how some of the ideas
introduced early in the book can be put into practice. In Chapter 7 we look at how more sophisticated approaches can be
used to develop true three-dimensional geometric modellers.

Introduction to surface and solid modelling

Introduction
A geometric modelling system is one which attempts to
manipulate and display information about three-dimensional
solid objects. There are various ways in which such objects
can be held within a computer system and in this chapter we
review some of the current approaches and try to indicate
some of the underlying techniques. These are often exten-
sions of ideas that have been discussed in previous chapters.

Several types of modeller are now recognized. These hold
information about the components being modelled to a
greater or lesser extent. For some applications it is not
necessary to have a complete description of the part. In such
cases it is sufficient to use a simpler (and hence less expensive)
modelling system. When a precise complete description of the
component is needed very often a full solid modeller is used.
The various types of system available are discussed in the
next section.

Types of geometric modeller
It is usual to classify geometric modellers as being of one of
three types. It should be borne in mind however that these
types are broad and the divisions between them are by no
means clear cut. Any given system may have aspects of all
three. To some extent the three types represent increasing
levels of sophistication in the modeller and what it is capable
of representing. The types are:

— wireframe modellers
— surface modellers
— solid modellers

We have met the idea of a wireframe modeller before. Such a
system stores the edges of a component together with defining

points for those edges. Thus when it is displayed the image produced is like a collection of wires in space marking the boundaries of the part. This is the simplest type of modeller to develop and clearly it does not represent a complete description of the geometry involved. Nonetheless, for applications where the parts being designed are simple such a modeller is useful.

It is usually claimed that there is ambiguity inherent in a wireframe model. For example, Figure 7.1 shows a wireframe model of a matchbox (a cuboid). It is not clear, however,

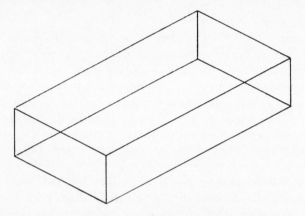

Figure 7.1 *A wireframe cuboid without hidden line removal*

whether we are looking at the tray, the cover or the assembly of the two; Figure 7.2 shows these possible interpretations with the appropriate edges removed to provide better visualizations.

This type of interpretation could be carried out by the system if it had information about where the faces of the cuboid lay. This brings us on to surface modellers. Here a user is allowed to introduce surface patches, usually to regions bounded by edges of the wireframe model. Thus more information is provided to the system. These surface patches may be simply planar ones. Some CAD systems (for example for architectural design) comprise a wireframe modeller with the ability to add in faces to provide the facility to describe and display unambiguously simple geometric shapes.

Normally a surface modeller also allows more complicated

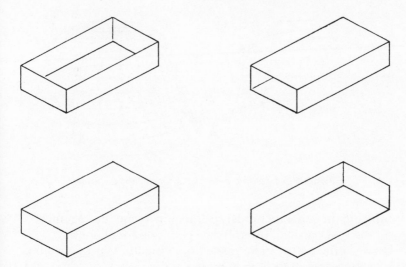

Figure 7.2 *Possible interpretations of a wireframe cuboid*

surface forms, often in terms of the Bézier and/or B-spline formulations. Such systems usually do not require the user to have a detailed knowledge of how the surfaces are described internally. For example, axi-symmetric surface patches are often obtained by rotating a plane curve about a specified axis to produce a surface of revolution. An alternative approach is to have the user specify the boundary of the region to be occupied by the patch and then let the system fill in the surface for itself. Figure 7.3 shows the construction of a surface model of a truncated cone given the original wireframe description also shown. Care is needed in such cases. The interpolation algorithm used by a CAD system is predefined and may not necessarily produce the patch the user had in mind.

Surface modellers by definition hold more information than wireframe modellers. However they do not necessarily hold all possible geometric data, nor do they need to. A typical application is in the design and manufacture of press tooling for producing complex shapes. The surfaces of the punch and die need to be described accurately as surfaces but a wireframe model suffices for the main body of the punch itself or for the block in which the die is sunk. Only over a restricted area is the more detailed description required. By

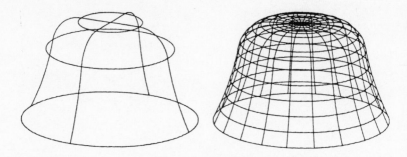

Figure 7.3 *Example construction of a surface model*

bringing the separate models of the punch and die together on the CAD system, interference checking for clashes between the parts can be carried out visually. If necessary the user can zoom in to look in more detail at any area of especial concern and the inclusion of surface patches means that the geometric detail is there to be inspected. Alternatively sections (usually plane) can be taken through the parts to ease the burden of visual checking by reducing the number of spatial dimensions from three to two. The fact that patches are present in the model means that there is indeed something to section and the CAD system runs through a surface − surface (or plane − surface) intersection routine to generate the required intersection curves.

In this way surface modellers hold more geometric details of complicated parts in those areas of critical concern to the design. The extra information can be used in many ways. As described above, some clash detection can be undertaken. Surface areas can be calculated where patches exist. These can be used, for example, to provide a quick estimate of the volumes of thin-walled vessels or to assess the rate of cooling of plastic material in an injection moulding process.

However the description held is not complete and surface modellers have their limitations. The surface definition may not be automatic and it can be time-consuming manually to insert surface patches at all areas of interest. This activity also requires a certain familiarity with the ways in which the patches are defined; for example it may be necessary to understand how control points affect the surface shape (as discussed in Chapter 5). A further drawback is that such properties as volumes and centres of gravity cannot normally be found due to the lack of the full geometry. For applications in which the surface geometry is inherently complex, any computer-aided approach is bound to be stretched to its limits, and a surface modeller (or good B-rep solid modeller as described in the next section) often proves to provide the best solution.

Wireframe and surface modellers do not attempt to retain a complete geometric description of a component. Solid modellers, on the other hand, aim to do precisely this. As the name suggests, they hold sufficient data to be able to tell where the solid material of a part lies and where it does not. As a consequence volumes and such other properties as centres of gravity and moments of inertia are easily obtainable. Indeed some systems hold this information all the time regardless of whether the user has requested it or not. Solid modelling is discussed in more detail in the next section.

Solid modelling

Two approaches to solid modelling are usually identified. These are:

— constructive solid geometry (CSG)
— boundary representation (B-rep)

In fact these refer more to how the geometric information is stored within the system than to how a user perceives what is happening. We start by considering the CSG approach as this is also a means for allowing the user to describe objects to the system. MODCON, discussed in the last chapter, uses an elementary form of CSG representation for its user interface.

In CSG input, the user builds up the parts he wishes to design out of simpler shapes using Boolean set theoretic operations. These operations are union, intersection and difference, although the terminology does vary from system to system. Figure 7.4 shows two cuboidal blocks; the effects of applying the Boolean operations are shown in Figure 7.5. The union of two parts is the combined volume occupied by both, the intersection is the region common to both, and there are two possible differences obtained by subtracting the material of one of the parts from the other.

To start the construction process the modeller usually provides a number of standard primitive shapes. These

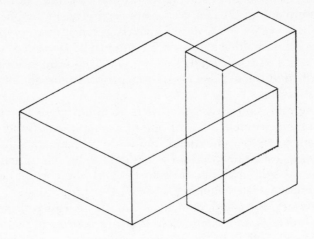

Figure 7.4 *Two solid shapes*

include such items as blocks, cones and spheres. The primitives are then used as building blocks and are combined by the Boolean operations to form more complicated shapes. These in turn can be stored and combined with other shapes to produce complete components.

A solid modeller is truly a CSG if it uses the description of a part as a Boolean combination of primitives as its means for storing geometric information. This is a very compact way of holding data, but effectively the part needs to be regenerated every time it is accessed. In the B-rep approach the user's description is translated so that the complete part is held as a collection of surfaces which represent the boundary faces of the part. In this way a B-rep solid modeller is a full surface model description of the geometry and the split between surface and solid modellers starts to become blurred.

B-rep modellers may still allow the user to input data in CSG form and indeed this is a natural way for the user to interact with the system. The point is that these data are turned into surface information internally. Thus the amount of data that has to be stored is large but it is capable of being manipulated easily. Furthermore the user can carry out operations that the plain CSG approach would not permit. For example the user may form the union of a cylinder and a rectangular block, a typical CSG-union operation. If he then wants to

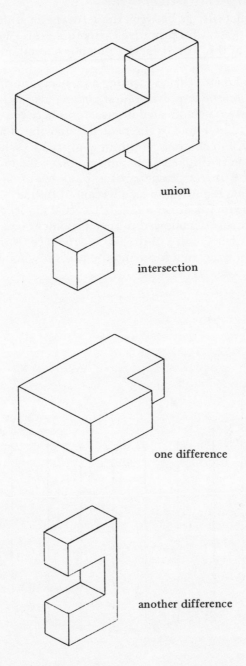

union

intersection

one difference

another difference

Figure 7.5 *Boolean operations for combining solids*

form a fillet surface between these two parts at their join it is unlikely that a CSG primitive exists to do this. However, the addition of this sort of surface patch is natural in a B-rep system.

As a solid modeller is carrying a vast amount of geometric information the response to user commands can be slow, particularly if complicated intersections between parts are required to be found or if repeated redrawing of a part is necessary. For this reason a third approach to solid modelling is sometimes introduced, the idea of a facetted modeller.

Here the surface geometry of a component is approximated by breaking it down into a collection of (usually) planar facets. Figure 7.6 shows a simple facetted model of a cylinder. Usually the user is allowed to specify some tolerance to control either the size of the facets produced or the error between the facetted form and the true one. Because the

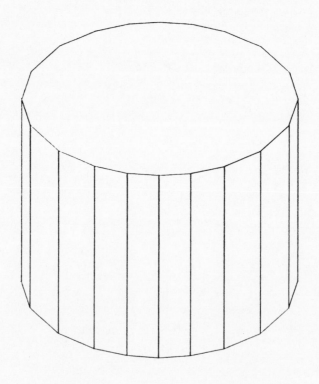

Figure 7.6 *A facetted circular cylinder*

modified geometry comprises much simpler entities (even if there are more of them), computations for intersections and so on are speeded up. The user needs to bear in mind, however, that it is an approximation that is being used and information such as dimensions and volumetric properties is only as accurate as the tolerance on the facetting allows.

Of the two basic types of solid modeller, the pure CSG approach is good and sufficient for parts which are composed of simple geometric entities. It is natural for the human mind to analyse and decompose reasonably complex objects into their more simple, geometrically regular constituents. A CSG system allows the reverse process of synthesis. The connecting rod example of Chapter 6 is one that it is natural to apply CSG to. As seen there, four solid and two cavity primitive shapes are sufficient to define the component. It is more difficult to add fillet and other blending radii to the model. However, if these are not critical, they can be omitted at the design stage and introduced during manufacture as a consequence of the finite tool dimensions. Should something more precise be required, access to surface patches needs to be available. It is here that B-rep modellers have the advantage. They can be provided with a CSG-type interface for the user to define the initial form (the input being converted to provide a boundary representation of the component) and subsequently the user can be allowed to refine the design by selective insertion and/or modification of surface patches. For this reason, B-rep modellers are more common among commercial CAD systems.

We have thus identified three approaches to solid modelling: CSG, B-rep and facetted modellers. To some extent these can be misleading and not helpful to the user of such a system. They refer more to the way in which geometric information is handled within a system than to the way in which the user interacts with it. They are not mutually exclusive since it is possible to hold and manipulate the data in more than one form (provided that sufficient care is taken to ensure that discrepancies between different versions of the data do not creep in). For example, one could produce a modelling system which used the CSG approach for its user interface. Once it is input, the system could convert the information into B-rep form by finding the bounding surfaces of the described component, and could also hold an approximate facetted version of this to speed up display functions. Thus the user would be seeing a facetted model but the system would also be retaining the precise B-rep geometry to ensure that accuracy was not lost.

Obtaining volumetric properties

Since a solid model contains a complete description of the geometry of a component, it is possible, as noted before, to use it to obtain additional information. This might include such data as the volume of the part, the position of its centre of gravity and its moments and products of inertia about specified axes.

These quantities can be defined easily in terms of volumetric integrals over the complete body and indeed it would be possible to evaluate them in this way using the boundary information held in a B-rep system. However it is easier to split the component down into simpler shapes, perform the relevant computation for each one and then combine the results. The use of a facetted model is usually an advantage here as the simpler shapes used can be triangles and tetrahedra which are straightforward to deal with.

We illustrate the basic approach involved by considering the determination of the area enclosed by a plane closed curve. Such a curve is shown in Figure 7.7, together with a

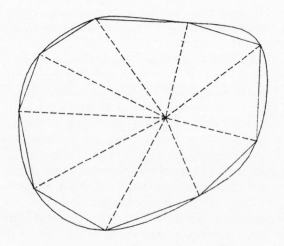

Figure 7.7 *Area by subdivision into triangles*

facetted approximation to it, that is, an inscribed polygon.
By selecting a point within the polygon and joining it to each
vertex, we divide the area into a number of triangles. If a
triangle in the xy-plane has vertices at the points (x_1, y_1),
(x_2, y_2) and (x_3, y_3), then its area is given by

$$\text{area} = \frac{1}{2} \begin{vmatrix} 1 & x_1 & y_1 \\ 1 & x_2 & y_2 \\ 1 & x_3 & y_3 \end{vmatrix}$$

$$= [x_1y_2 + x_2y_3 + x_3y_1 - x_2y_1 - x_3y_2 - x_1y_3]/2$$

This formula has the property that it gives a positive area if
the three vertices are given in anticlockwise order round the
perimeter, and a negative value if they are in clockwise order.
This means that the extra point we select does not have to be
inside the polygon. If the point is selected outside (as in
Figure 7.8), then some of the triangles are found to have
negative areas and these partly cancel in the summation to
find the area of the polygon. Thus our algorithm to find areas
need not check where the additional point is relative to the
area required. (However it should not be a large distance away
as loss of significant figures in the calculation may result.)

The above procedure extends to three-dimensional regions
bounded by surface patches; there the subdivision is into

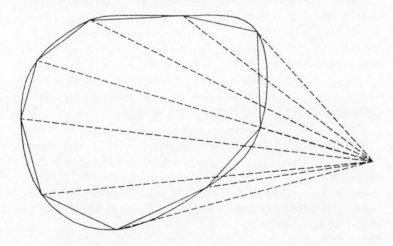

Figure 7.8 *Subdivision with vertex point outside the curve*

tetrahedra and the total volume can be found in a similar way.
In each case, rather than approximate the bounding curve
segment or surface patch by straight line segments or planar
facets, one could attempt to calculate the precise area or
volume involved. This requires knowledge of the parametric
form used and results in an integral which needs to be
evaluated. Except in very special cases this cannot be done
analytically. Numerical methods have to be applied and are
essentially equivalent to the use of the original facetting
procedure and so little, if anything, is gained.

Defining primitive volumetric shapes

Particularly with the CSG type of input, it is necessary for
the user to have access to a number of basic primitive shapes
which can be combined to build up the required complete
design. One way of providing these is to include a number of
standard shapes within the CAD system. These should be
parametrized in the sense that the user can change the sizes
of some or all of the dimensions.

However, this approach can limit the applicability of the
system. It is excellent for those components which can
naturally be subdivided into the primitive shapes of the
system, but if a particular shape is not immediately available,
it may make the component awkward to deal with. So
systems often also allow the user more freedom in defining
volumetric shapes. For a B-rep based system one way to
do this is to allow the user to create his own primitives
by starting with a wireframe model which is then surfaced
and declared to the system to be a solid entity. This type
of approach is certainly very flexible but it can be time-
consuming for the user and may demand detailed knowledge
of how the surfaces are defined internally.

One way of making this definition process semi-automatic
is by means of a technique known in some CAD systems as
'extrusion'. Here a plane curve is defined by the user with
standard two-dimensional construction commands. This
curve may be composed of straight lines and circular arcs or
more generally of Bézier or B-spline segments. This composite
curve is then copied by projecting it into the third dimension
through a distance specified by the user. Corresponding
points on the two curves are then joined and a ruled surface
between them is formed. If the original curve is closed, then a

cylinder results; by inserting planar surface patches to cover
the ends, a solid entity has been produced (Figure 7.9).

If, instead of extruding the plane curve along a straight
line, we rotate it about an axis (usually in its own plane),
then a simple surface of revolution is formed and as before
this can subsequently be treated by the solid modeller as a

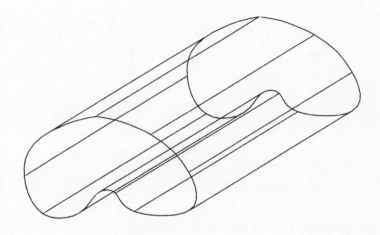

Figure 7.9 *General cylinder shape produced by extrusion*

solid, a solid of revolution. More generally we can move the
original plane curve along a more general curve in space.
This processs is known in some CAD systems as 'sweeping'.
It is a construction which is the same as that underlying
the DUCT system described in Chapter 5. For example
by moving a (small) circle along a curve in space we generate the shape of a pipe (Figure 7.10). Figure 6.14 shows
another example where a non-closed curve is swept round
a closed curve to obtain the bounding surface of a solid
entity.

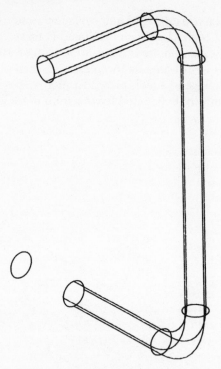

Figure 7.10 *Pipe produced by sweeping*

Hidden line removal and surface shading

With a geometric modeller we are producing a three-dimensional computer model of the component being designed. The crudest way of displaying this on the graphics screen is as a wireframe model, showing just the edges of all the faces that compose the part. With a little enhancement this is sufficient to be able to produce the usual plans and elevations that comprise a traditional engineering drawing. However in this process we are essentially producing a set of two-dimensional views and some information is being lost. CAD systems try to provide means of viewing the component as a three-dimensional (solid) entity. In this section we review some of the techniques that are used to provide this type of visualization on the two-dimensional graphics screen or plotter.

We consider firstly the case of displaying a component as a wireframe model. There is a danger that the picture produced

seems to consist of a vast collection of lines and curves and
the viewer has no idea of depth. The use of perspective can
help here and as this can be incorporated as part of the
viewing transformation it represents no large additional
computational overhead (cf. Chapter 2). An alternative
inexpensive method is the use of 'intensity-cueing'. The CAD
system knows how far away from the observer any particular
part of the displayed object is. By drawing distant parts using
lines of less intensity than those near the observer an illusion
of depth can be created.

Both these techniques rely directly on information already
available within the CAD system and are relatively easy to
implement. They are used on the newer breeds of work-
station which allow the image of the three-dimensional model
to be translated and rotated on the screen in real-time. The
ability to be able to move the model at will gives a dramatic
feel for the shape of the component even though it is actually
only displayed as a wireframe form.

A computationally more expensive way of enhancing a
wireframe picture of an object is by the use of hidden line
removal. Here edges that would not be visible to the observer
because they are behind other parts of the object are not
displayed. Figure 7.2 shows simple examples of hidden line
removal. Algorithms here are complicated, although they
are simplifed to some extent if the model is assumed to be a
facetted one, where the faces are all planar and are easier to
deal with.

A crude method of hidden line removal is to examine the
outward pointing normal to each face of the model. These
are straightforward to compute when the faces are planar.
An example for a cuboid is shown in Figure 7.11. If after
the model has been transformed to the viewing configuration,
the normal to a face points away from the observer then that
face is assumed to be hidden and it is not drawn. Only the
edges for faces whose normals have a component pointing
towards the viewer are displayed. This technique is not
foolproof; it is sure to work for solid components (without
holes) which are convex. If the object is re-entrant it can fail.
However this approach can be used to give some impression
of depth and it is useful as a quick means of eliminating some
faces from consideration before more sophisticated methods
are used.

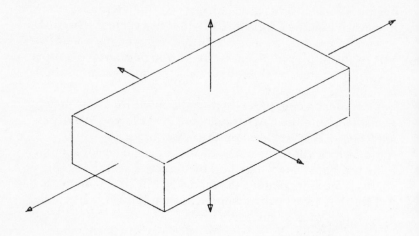

Figure 7.11 *Outward pointing normals*

To obtain a true hidden line display it is ultimately
necessary to compare many pairs of faces in the model to see
if one obscures the other. In the majority of cases a pair of
faces are well separated and do not overlap when seen from
the observer's position. A simple box-test of the extents of
the two faces will eliminate this case easily (as discussed in
Chapter 4). If the boxes round the faces do overlap we need
to investigate further. It is possible that the faces do not
interfere. If they do we need to consider the distances from
the observer. If one face is completely behind the other then
it is obscured and is not displayed. If one is partially hidden
we need to calculate intersections in order to be able to
display only that part or parts that are showing. In a more
general context it would be necessary to consider the case of
one face passing through the other so that they mutually
obscured parts of each other; for a B-rep solid model this
should not happen as the outer boundary is already known

and indeed the calculations for intersections have been
carried out previously. Once one face has been compared
against all others, if any part is found to be visible then it
can be plotted. As can be seen, hidden line removal is time-
consuming and for this reason a user may use it only infre-
quently as a means for checking the validity of his model.
For a review of methods for hidden line removal, the reader
is referred to Sutherland *et al.* (1974).

Moving away from wireframe displays of a geometric
model, we come to the technique of producing shaded image
pictures of the component. The aim is to provide a picture of
the model which is as realistic as possible. The usual method
is to use colour to fill in the surfaces of the solid model. In
order to select the intensity, it is assumed that the object is
illuminated by light from some point source. If we know the
direction of the surface normal at a point and the direction
to the light source, then various formulae can be used to
determine how much light would be reflected towards the
observer. Again the use of a facetted modeller is an advantage
since it eases some of the problems of determining the normals to the surfaces involved. To give a more realistic picture
it is usual to assume that there is general ambient lighting as
well as the particular point source. If the point source is
along the light of sight of the observer, then there is no need
to take account of shadows that might be cast by various
parts of the object on other parts.

As well as deciding how to shade a surface it is also necessary to determine how much of it is obscured by others in the
model. This can be done by comparing pairs of surfaces (or
facets) as for hidden line removal above. Alternatively, use
can be made of the graphics hardware. Loosely speaking, if
we display the surfaces starting with those further away from
the observer and moving closer then we can rely on the fact
that the later ones will automatically overwrite the earlier.
This approach needs to be refined to ensure it works every
time, unless we use it at the individual pixel level of the
graphics screen. At each pixel the surfaces present are identified together with their distance from the observer. The
closest is used and the pixel filled with the appropriate colour
and intensity.

A lot of sophisticated CAD systems allow the user to
perform surface shading, although the methods used may

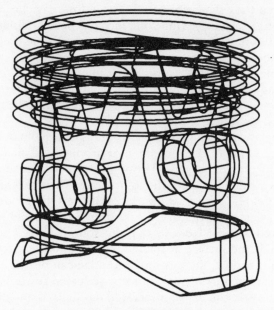

Figure 7.12 *Wireframe model (courtesy of Computervision Corporation)*

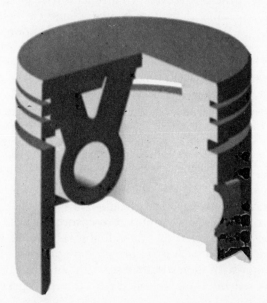

Figure 7.13 *Shaded solid image corresponding to the wireframe model shown in Figure 7.12 (courtesy of Computervision Corporation)*

vary from system to system. Figures 7.12 and 7.13 show a wireframe model and the corresponding shaded solid image.

This has been intended as a brief review of some of the methods used to enhance the representation of a solid model. For a more detailed discussion see, for example, Foley and Van Dam (1982). It needs to be borne in mind that sophisticated methods are expensive both in terms of the computational time involved and in the price of the relevant software. Using a facetted version of the model eases some of the difficulties involved. Indeed most current commercial CAD systems use some version of facetting as is sometimes obvious from the results produced.

From an engineering point of view it can be argued that the use of these techniques is not entirely relevant. They work by failing to display information such as hidden lines or surfaces. Often the user is as interested in those parts of his design which cannot be seen as in those which are visible as he needs to ensure these interact correctly. However, these enhanced representations are a means to check finally that an overall design is correct and, if appropriate, that it is aesthetically pleasing. They are also valuable for technical illustrations, publicity material and more generally when communicating with people not accustomed to reading engineering drawings.

References

Although this list of references in no way represents a complete bibliography of the CAD literature, it does contain items that are not referred to in the text and is intended to suggest background and further reading material.

Ayres, F. (1974) *Theory and Problems of Matrices*, McGraw-Hill.

Baer, A.; Eastman, C.; Henrion, M. (1979) Geometric modelling: a survey. *Comput. Aided Des.* 11, pp. 253-272.

Ball, A.A. (1984) Reparametrization and its application in computer-aided geometric design. *Int. J. Num. Meth. Eng.* 20, pp. 197-216.

Bézier, P. (1972) *Numerical Control, Mathematics and Applications*, Wiley, New York.

Böhm, W.; Farin, G.; Kahmann, J. (1984) A survey of curve and surface methods in CAGD. *Comp. Aided Geom. Des.* 1, pp. 1-60.

Burden, R.L.; Faires, J.D.; Reynolds, A.C. (1978) *Numerical Analysis*, 2nd ed., Prindle, Weber and Schmidt, Boston.

Chan, Y.K.; Knight, W.A. (1979) Computer-aided manufacture of forging dies by volume building. *J. Mech. Working Tech.*, 3, pp. 167-183.

Chan, Y.K.; Knight, W.A. (1980) MODCON: a system for the CAM of dies and moulds. *Proc. CAD80.*

Chan, Y.K.; Mullineux, G.; Knight, W.A. (1979) Progress in the computer-aided design and manufacture of hot forging dies. *Proc. 20th. Int. M.T.D.R. Conf.* pp. 29-38.

Coons, S.A. (1967) *Surfaces for Computer Aided Design of Space Forms*, Report MAC-TR-41, MIT

Cox, M.G. (1972) The numerical evaluation of B-splines. *J.I.M.A.* 10, pp. 134-149.

de Boor, C. (1972) On calculating B-splines. *J. Approx. Th.* 6, pp. 50-62.

de Boor, C. (1977) Package for calculating with B-splines. *SIAM J. Nu. Anal.* 14, pp. 441-472.

de Casteljau, F. (1959) *Outillage Methods Calcul* Andre Citroen Automobiles SA, Paris.

de Casteljau, F. (1985) *Shape Mathematics and CAD*, Kogan Page, London.

Denavit, J.; Hartenberg, R.S. (1955) A kinematic notation for lower-pair mechanisms based on matrices. *J. App. Mech.* pp. 215-221.

Faux, I.D.; Pratt, M.J. (1979) *Computational Geometry for Design and Manufacture*, Ellis Horwood, Chichester.

Ferguson, J.C. (1964) Multivariate curve interpolation. *J.A.C.M.* 11, pp. 221-228.

Foley, J.D.; Van Dam, A. (1982) *Fundamentals of Interactive Computer Graphics*, Addison-Wesley, Reading, Massachusetts

Forrest, A.R. (1972) On Coons' and other methods for the representation of curved surfaces. *Comp. Graph. and Image Proc.* 1, pp. 341-359.

Gordon, W.J.; Riesenfeld, R.F. (1974) B-spline curves and surfaces. In *Computer-Aided Geometric Design* (R.E. Barnhill and R.F. Riesenfeld, eds.), Academic Press.

Hammond, D.S. (1983) *The Representation of Complex Surfaces for the Production of Numerically Controlled Toolpaths*, Brunel University, London.

Lee, C.S.G. (1982) Robot arm kinematics, dynamics and control. *I.E.E.E. Computer*, pp. 62-80.

Medland, A.J. (1986) *The Computer-Based Design Process*, Kogan Page, London.

Mullineux, G.; Knight, W.A. (1980) Integrated CAD/CAM of dies for hot forgings. *Proc. 21st. Int. M.T.D.R. Conf.* pp. 593-602.

Mullineux, G. (1982a) Approximating shapes using parametrized curves. *I.M.A. J. App. Math.*, 29, pp. 203-220.

Mullineux, G. (1982b) Reducing the degree of high order parametrised curves. *Proc. CAD82 Conf.* pp. 303-314.

Mullineux, G. (1984) Surface fitting using boundary data. *Proc. CAD84 Conf.*

Newman, W.; Sproull, R. (1979) *Principles of Interactive Computer Graphics*, McGraw-Hill.

Scheid, F. (1968) *Theory and Problems of Numerical Analysis*, McGraw-Hill.

Sutherland, I.E.; Sproull, R.F.; Schumacker, R.A. (1974) A characterization of ten hidden-surface algorithms. *Comp Surv* 6, pp. 1-55.

Walsh, G.R. (1975) *Methods of Optimization*, Wiley, London.

Index